# Environmental Responses

edited by
Andrew Blowers and Steve Hinchliffe

**WILEY**

wiley.com

in association with

The Open
University

This publication forms part of an Open University course U216 *Environment: Change, Contest and Response*. The complete list of texts that make up this course can be found on the back cover. Details of this and other Open University courses can be obtained from the Course Information and Advice Centre, PO Box 724, The Open University, Milton Keynes MK7 6ZS, United Kingdom: tel. +44 (0) 1908 653231, e-mail general-enquiries@open.ac.uk

Alternatively, you may visit the Open University website at www.open.ac.uk where you can learn more about the wide range of courses and packs offered at all levels by The Open University.

To purchase a selection of Open University course materials visit the webshop at www.ouw.co.uk, or contact Open University Worldwide, Michael Young Building, Walton Hall, Milton Keynes MK7 6AA, United Kingdom for a brochure. tel. +44 (0) 1908 858785; fax +44 (0) 1908 858787; e-mail ouwenq@open.ac.uk

First published 2003 by John Wiley & Sons Ltd in association with The Open University

The Open University
Walton Hall
Milton Keynes
MK7 6AA
United Kingdom
www.open.ac.uk

John Wiley & Sons Ltd
The Atrium
Southern Gate
Chichester
PO19 8SQ

www.wileyeurope.com or www.wiley.com

Email (for orders and customer service enquiries): cs-books@wiley.co.uk

Other Wiley editorial offices: John Wiley & Sons Inc., 111 River Street, Hoboken, NJ 07030, USA; Jossey-Bass, 989 Market Street, San Francisco, CA 94103–1741, USA; Wiley-VCH Verlag GmbH, Boschstr. 12, D–69469 Weinheim, Germany; John Wiley & Sons Australia Ltd, 33 Park Road, Milton, Queensland 4064, Australia; John Wiley & Sons (Asia) Pte Ltd, 2 Clementi Loop #02–01, Jin Xing Distripark, Singapore 129809; John Wiley & Sons Canada Ltd, 22 Worcester Road, Etobicoke, Ontario, Canada M9W 1L1.

*Library of Congress Cataloging-in-Publication Data*

A catalogue record for this book is available from the Library of Congress.

*British Library Cataloguing in Publication Data*

A catalogue record for this book is available from the British Library.

ISBN 0 470 85005 1

Edited, designed and typeset by The Open University.

Printed by The Bath Press, Glasgow.

1.1

# Contents

# Series preface

*Environment: Change, Contest and Response* is a series of four books designed to introduce readers to many of the principal approaches and topics in contemporary environmental debate and study. The books form the central part of an Open University course, which shares its name with the series title. Each book is free-standing and can be used in a wide range of environmental science and geography courses and environmental studies in universities and colleges.

This series sets out ways of exploring environments; it provides the knowledge and skills that enable us to understand the variety and complexity of environmental issues and processes. The ideas and concepts presented in the series help to equip the reader to participate in debates and actions that are crucial to the well-being – indeed the survival – of environments.

The series takes as its common starting point the following. First, environments are socially and physically dynamic and are subject to competing definitions and interpretations. Second, environments change in ways that affect people, places, non-humans and habitats, but in ways that are likely to reflect differing degrees of vulnerability. Third, the unsettled, uncertain and uneven nature of environmental change poses significant challenges for political and scientific institutions.

The series is structured around the core themes of **changing** environments, **contested** environments and environmental **responses**. The first book in the series sets out these themes, and (as their titles indicate) each of the three remaining books takes one of the themes as its main concern.

In developing the series we have observed the rapidly changing world of environmental studies and policies. The series covers a range of topics from biodiversity to climate change, from wind farms to genetically modified organisms, from the critical role of mass media to the measurement of ecological footprints, and from plate tectonics to global markets. Each issue requires insights from a variety of disciplines in the natural sciences, social sciences, arts, technology and mathematics. The books contain chapters by a range of authors from different disciplines who share a commitment to finding common themes and approaches with which to enhance environmental learning. The chapters have been read and commented on by the multidisciplinary team. The result is a series of books that are unique in their degree of interdisciplinarity and complementarity.

It has not been possible to include all the topics that might find a place in a comprehensive coverage of environmental issues. We have chosen instead to use particular examples in order to develop a set of themes, concepts and questions that can be applied in a variety of contexts. It is our intention that as you read these books, individually or as a series, you will find thought-provoking and innovative approaches that will help you to make sense of the issues we cover, and many more besides. From the outset our aim has been to provide you with the equipment necessary for you to become a sceptical observer of, and participant in, environmental issues.

In the series we shall talk of both 'environment' and its plural, 'environments'. But, what do these terms mean? As a working definition we take 'environment' to indicate surroundings, including physical forms and features (land, water, sky) and living species and habitats.

'Environments' signifies spaces (areas, places, ecosystems) that vary in scale and are connected to, are a part of, a global environment. For instance, we may think in terms of different environments, such as the Scottish Highlands, the Sahara Desert or the Siberian taiga (forest). We may think in terms of components of environments, for instance housing, industry or infrastructures of the built environment. The phrase 'natural environments' may be used to emphasize the contribution of non-human organisms and biological, chemical and physical processes to the world's environment.

It is useful to keep in mind four defining characteristics of environment: environment implies surroundings; nature and human society are not separate but interactive; environment relates to both places and processes; and environments are constantly changing. These characteristics are merely a starting point; as we proceed we shall gradually reveal the relationships and interactions that shape environments.

There are several features of the books that are worthy of comment. While each book is self-contained, you will find references to earlier and later material in the series depicted in bold type. Some of the cross-references will also be highlighted in the margins of the text so that you may easily see their relevance to the topic on which you are currently engaged. The margins are also used in places to emphasize terms that are defined for the first time.

Another feature of the books is the interactivity of the writing. You will find questions and activities throughout the chapters. These are included to help you to think about the materials you are studying, to check your learning and understanding, and in some cases to apply what you have learned more widely to issues that arise outside the text. A final feature of the chapters is the summaries that appear at the end of each major section, to help you check that you have understood the main issues that are being discussed.

We wish to thank all the colleagues who have made this series and the Open University course possible. The complete list of names of those responsible for the course appear on an earlier page. Particular thanks go to: our external assessor, Professor Kerry Turner; our editors, Melanie Bayley, Alison Edwards and Lynne Slocombe; our tutor panel, Claire Appleby, Ian Coates and Arwyn Harris; and to the secretaries in the Geography Discipline, Michelle Marsh, Jan Smith, Neeru Thakrar and Susie Hooley, who have all helped in the preparation of the course. Last but not least, our thanks to our course manager, Varrie Scott, whose efficiency and unfailing good humour ensured that the whole project was brought together so successfully.

Andrew Blowers and Steve Hinchliffe
Co-chairs of The Open University Course Team

# Introduction

Andrew Blowers and Steve Hinchliffe

## What are environmental responses?

To respond often means to reply or answer. An environmental response might therefore be thought of as people responding to environmental problems, challenges or questions. This is what we mean by the term *responses to* environments. For example, a response to the perceived threat of global climate change might be to reduce our global fossil fuel consumption. There are, of course, other possible responses, so there is likely to be a good deal of discussion and weighing up of the costs and benefits of a variety of environmental responses.

In order to respond effectively to environments it is necessary to understand how non-human aspects of environments are themselves adjusting to change. This kind of discussion leads us on to thinking about environmental responses in a second way which we can term *responses of* environments. To illustrate this, we can continue with the same example. In order to judge the effectiveness on climate processes of limiting the amount of fossil fuels consumed we would need to understand how climate processes would respond to such a reduction in $CO_2$ emissions. On the one hand, it may be that the world's environments would hardly notice the reduction, and would carry on responding to the last few centuries of increased atmospheric carbon concentrations. In this case the responses of environments may be so slow that the human response is too little and too late. On the other hand, environments may be much more responsive, and any social response, quickly implemented, could have a huge effect on future climates.

In truth, these are difficult issues, and a lot of work is needed in order to understand how environments respond. What is important for our purposes is to note that environmental responses include the responses *of* environments as well as *to* environments. Moreover, as you read the chapters in the book, note how these responses are often interrelated; only by understanding how physical and social systems respond to change can further responses – in the form of policies, management or other forms of action – be developed. To respond effectively requires an interdisciplinary approach to environmental issues. Our intention is to combine the insights of individual disciplines with an integrated, interdisciplinary approach to the study of environmental responses.

People respond to environments in a number of ways that range from doing very little, or waiting to see what happens, to attempting to take charge of any situation.

In this book you will meet a number of examples that you may want to position between these two extremes. For now we can trace three forms of response.

inactive response

First, at the most passive level, response may be described as **inactive**. This may be brought about by indifference to environmental deterioration or the optimistic belief that the problem will go away. It may result from an inability through lack of power or resources to do anything about it. It may reflect a recognition that it is impossible to avoid changes that are unforeseeable or beyond control – for instance, the fatalistic acceptance of a gradually deteriorating neighbourhood brought about by poverty and powerlessness. In its most abject form inaction may lead to elimination or extinction of communities or ways of living.

reactive response

A second category of response may be called **reactive**, describing those forms of action precipitated by any change in environmental circumstances. Reacting to a marine oil spill, for example, often involves actions to stem the spread of the pollutant and stop it reaching the shallow inland waters. This is an example where the change is sudden and discernible. Other reactive responses tend to be quite passive, the gradual adjustment to more slowly changing conditions. For example, it may be possible to adjust to sea level rise for a while by building higher defences but, at some point, more drastic action such as managed retreat may become inevitable, a form of reactive adaptation to inevitable inundation.

proactive response

Where action is planned in anticipation of environmental change it comes in the third category, **proactive response**. This may be a response to a known or predicted change, trend or event. For example, predictions of considerable growth in households in southern England leads to the plan for the location of major centres of development around Milton Keynes, the Thames Gateway, the M11 and around Ashford in Kent. Forecasts of up to three times the amount of air travel in the UK over the next thirty years require an airport strategy that attempts to reconcile competing demands for economic growth and environmental conservation. These predictions are not always agreed upon (though often presented as inevitable).

See **Bingham and Blackmore (2003)** for a discussion of the precautionary approach to environmental issues.

Predicting the future is extremely fraught, yet it is often wise to plan for environmental eventualities that are not easy to foresee or predict and which may or may not occur. It is often necessary to take precautionary action now, even in the absence of scientific certainty, in order to prevent the possibility of future environmental damage. Even though we know so little about the future consequences of climate change, it is often argued that it makes economic and scientific sense to act now, knowing that we might be wrong, rather than adopt a fatalistic stance and risk the worst possible outcomes. To understand more of these complex issues, it is useful to understand uncertainties and risks.

# Uncertainties and risks

Responses of whatever kind require us to consider seriously the problem of not knowing what the future may hold (uncertainty) and of acting in conditions of uncertainty (taking risks). For these reasons, risk and uncertainty form the main analytical concepts that recur throughout the chapters in this book.

In making any response we must undertake, consciously or not, a **risk assessment**. Formally, a risk assessment involves a calculation of probability, the chance of an event (usually but not necessarily a hazard) occurring at some future point in time. It may not always be possible to provide a numerical probability of a risk occurring. Indeed, the further ahead we look the wider the range of probabilities and the more uncertain risk assessment becomes. For example, it is difficult to predict all the changes that might impact on a radioactive waste repository over a period of 10,000 let alone 100,000 years and it is impossible to assume social stability even over the life of a single generation. These issues of long-term timescales and risk are investigated in Chapter One. The perception of risk adds further complications. It is the case that some risks instil fear or dread while others are treated with surprising disdain. Classic instances of dread risk are the fear of flying or of radioactivity, where there are low probabilities but high consequences in the event of accident. By contrast there is the widespread indulgence in smoking where the probabilities of disease and death are high. Various explanations for these seeming contradictions have been given, involving individual control over risk avoidance, trust in institutions and personal choice. These matters are discussed in Chapter Two in relation to problems of managing urban environments.

*risk assessment*

*See also **Bingham and Blackmore (2003)** on risks and uncertainties.*

Environmental **risk management** – applying proactive methods to reduce or minimize risk – is today a commonplace activity in business, public services and international institutions. It involves policies, protocols and procedures often backed by incentives or sanctions. It often requires choices to be made between alternative courses of action, each carrying different costs and consequences (choices are often linked to economic analyses; see Chapter Three). But risk assessment and risk management are activities surrounded by uncertainty. It is very often the case that we do not know how environments will respond. To this scientific uncertainty must be added considerable political uncertainty about the way individuals, institutions and societies will respond to a change in policy (see Chapter Four). For example, we might pass legislation to make cars more efficient in terms of their petrol consumption, but if people respond by driving further, faster and in bigger cars the legislation will prove less effective than intended.

*risk management*

Beyond the uncertainties of people's consumer behaviour, there are significant uncertainties associated with the required level of social and international organization that many transboundary environmental issues require. Transboundary processes have stimulated transnational responses in the form of international regimes such as the United Nations Framework Convention on Climate Change, Convention on Biodiversity and the Convention to Combat Desertification. Nonetheless, these regimes are not autonomous but must rely for implementation on the cooperation of the individual nation states that have signed up to them. Cooperation is likely to prove difficult in conditions of uneven development, resulting in persistent environmental inequalities. (See Chapters Five and Six on global climate change and biodiversity conservation for issues of this kind.)

Responses occurring now have implications for the future – in some cases, such as managing radioactive materials, for the unimaginably far future. Taking care not to bequeath harm to future generations and to ensure that they can meet their needs are the principles of sustainable development. Interpreting sustainable development is a major challenge since it implies that we should be able to envisage the kinds of future that are desirable and that we can also provide the means to achieve them. Sustainable development requires not only agreement on means and ends but a willingness on the part of the present generation to forego desirable activities that may impose burdens on future generations.

# How can we respond?

The chapters in this book ask you questions about how environmental responses are to be understood and acted upon. In doing so, you should bear in mind the general point being made in this introduction, that to act, or not act, requires taking risks in conditions that are far from being certain. The key questions asked in the chapters are as follows:

- What tools and approaches are available for managing environmental problems and how useful are they?
- What are the constraints and opportunities for sustainable development?
- Can environmental inequalities be reduced?
- What kinds of environmental futures are desirable and how might we choose among them?

Thinking about these questions is a way in to trying to understand and to influence environmental responses.

The first question on approaches to managing environmental problems is answered in different ways in each chapter. The first chapter considers responses to radioactivity in Cumbria in north-west England. It finds that there are uncertainties over scientific evidence and predictions leading to contests among scientists and between scientists and lay people who may introduce knowledge and expertize ignored by official experts. The chapter argues that there is a need to open up participation in decision making on issues involving risks to environments in order to expose a wider range of relevant knowledge. Doing so allows for more open and frank discussions of what it means to live with uncertainty and risk. The approach suggests that we may need to re-think the relationships between science and wider society in this risky and uncertain world.

*technological response*

The following three chapters each focus on specific types of environmental response. Chapter Two is concerned with **technological responses**. These include methods of energy conservation, pollution abatement, waste reduction and transport and urban planning. Among the methods advocated are: combined heat and power; use of renewable energies; life-cycle analysis; and reuse and recycling of waste materials. These environmental technologies combined with urban planning provide for the practical implementation of policies of sustainable development, particularly with respect to climate change and clean air.

The tools and approaches that may be used to encourage or enforce environmental policies are the subject of Chapter Three, which provides an **economic analysis**. The chapter moves from comparing different economic approaches to environmental management, to asking searching questions about the role of states in setting their own environmental standards, to the overall aims of economic growth. The chapter ends by asking whether or not responding within the paradigm set by current economic aims misses the point that there are other ways of encouraging more sustainable and more enjoyable lifestyles.

economic analysis

Environmental politics is the subject of Chapter Four. The argument here is that **political response** involves the use of power, which is unevenly distributed in society. Echoing the first chapter, effective response must be acceptable, must be seen to be accountable and its outcomes must be seen to be fair. These issues are illustrated through two comparative case studies, which show how political opportunities to participate depend on the political capacity of those involved and on the openness of political systems. Although there is evidence of increasing opportunity to participate, unequal power relations persist and strongly influence the eventual outcomes of environmental controversies.

political response

Unequal power is an underlying reason for conflict over environmental policy making at an international level and Chapter Five takes up the theme of political response in the context of climate change negotiations. While there is broad scientific consensus on climate change processes, there is disagreement over the degree of change and the distribution of causes and impacts. Scientific uncertainty is compounded by the political uncertainties created by the diversity of interests of different states and particularly between the North and the South. The unevenness of the risks and the differential capacities to respond encourage disagreement. However, at least it may be said that international effort is being made to face up to a common threat.

The constraints and opportunities for response are the subject of the final chapter. Taking biodiversity as its focus it considers how measurement of biodiversity and the application of economic valuation to species and habitats can encourage conservation, at least in the short term. Not surprisingly, the idea that it is possible to achieve both economic growth and ecological conservation has been promoted by policy makers because it suggests that tough choices do not have to be made. But, as the later sections of the chapter show, there is another side, the paradox that biotechnology, which brings prosperity in the short term, can ultimately destroy the very biodiversity on which it depends.

Throughout the book there is evidence that can address the second key question: what are the constraints and opportunities for sustainable development? The technological potential for conservation emphasized in Chapter Two combined with the political commitment to ecological modernization set out in Chapters Three, Four and Six suggest that the opportunities exist to ensure a sustainable environment. In contrast, the uncertainties associated with much of modern technology (Chapters One and Six) allied to a tendency for market values to predominate in a political system with limited participation (Chapters Three and Four) indicate that the constraints may prove overwhelming in the long run.

Above all, the inequalities of power that are exposed between experts and public (Chapter One), between economic and environmental interests (Chapter Four) and between North and South (Chapters Five and Six) present formidable barriers to the political responses that are necessary to protect environments. As long as the impacts of environmental change are uneven, no matter how severe for some environments, the prospects for global agreements will be low. But, as Chapter Five shows, there are already indications of a sense of common purpose to protect a common inheritance. Whether it is possible to turn this into workable policies and successful implementation will depend on there being a positive answer to the third key question: can environmental inequalities be reduced?

In reading these chapters you will realize that environmental responses are challenging. They require us to look seriously at the way we normally go about things. Does science give us the necessary answers? Can we trust markets to guide our actions? Are conventional forms of politics reliable in giving us good decisions that take ever larger timeframes and areas into account? It is easy to be negative and regard the world's environments as doomed. But in asking the questions that we have set out in this book, we are also throwing out a challenge to the conventional ways of approaching environmental issues. In engaging with the chapters in this book it is hoped that the challenge will be understood more coherently and will be taken one step further.

The chapters add up to a series of resources for thinking about environmental futures, the possibilities and challenges. Evidently, though, we are at an early stage in recognizing our collective responsibilities and in setting out what needs to be done regarding environmental responses. It is clear that we have reached the point at which we must begin to address the final question for this book: what kinds of environmental futures are desirable and how might we choose among them? You will find hints to answers to this question as you read through the chapters, so it will be worth keeping a note as you work through the book. Our thoughts on this challenging question are given in the Epilogue.

# Reference

Bingham, N. and Blackmore R. (2003) 'What to do? How risk and uncertainty affect environmental responses' in Hinchliffe, S.J., Blowers, A.T. and Freeland, J.R. (eds) *Understanding Environmental Issues*, Chichester, John Wiley & Sons/The Open University (Book 1 in this series).

# Chapter One

# Environmental responses: radioactive risks and uncertainty

Steve Hinchliffe and Andrew Blowers

## Contents

# 1    Introduction

See **Morris et al. (2003)** for discussion of the causes of environmental change.

See **Bingham and Blackmore (2003)** for a discussion of risks, hazards and uncertainties: *risks* are defined as the probabilities of a known outcome, or hazard, occurring; *uncertainties* refer to the unknown and unknowable aspects of a situation.

Environments change constantly, often imperceptibly, and often without much cause for concern. Yet in some circumstances environmental changes produce anxiety. There are concerns over what is happening and how will it affect people, animals, plants and environments more generally. In other words, what hazards are being created and what are the risks? In order to understand the responses to these kinds of question we need to understand risks and uncertainties. We also need to understand how risks and uncertainties are handled in socially, politically and scientifically complex societies.

There are plenty of examples that could be used to illustrate how environments can change in ways that make them more hazardous. Global shifts in the Earth's energy flows, brought about in the main part by anthropogenic emissions of greenhouse gases, may, many argue, increase the risks that some places, plants and animals face as a result of accelerated climate change (see **Blackmore and Barratt, 2003**). Similarly, the introduction of genetically modified organisms may, some argue, pose a health hazard for other living organisms and a health as well as an economic hazard for certain groups of people (see **Bingham, 2003**). In this chapter we take a particular issue – that of radioactive contamination – in order to open up a general discussion of responses to potentially hazardous events. In using the word 'responses' we mean two things. First, we are referring to the responses *of* environments to change. These, we will argue, are complex and characterized by a range of physical and social activities. Second, we are referring to the active responses that people make in order to adjust *to* these wider changes. These can often be described as policy responses. The distinction is in part one of convenience and we envisage the responses of environments and the responses to environments to be interlinked.

Our setting for the chapter is Cumbria in north-west England: see Figure 1.1(a). Cumbria is well known as a tourist destination based on the attraction of the Lake District, but parts of Cumbria are also associated with hazards that are linked to the possibility of high levels of radioactive substances in the local environment. In part this hazard relates to the presence of a nuclear reprocessing plant on the west coast of the county at Sellafield (formerly known as Windscale) **(Blowers and Elliott, 2003)**. The plant has been plagued by controversy since it opened in the 1950s. Its discharges of radioactive material into the Irish Sea have produced disagreements between UK and Irish governments over the acceptability of the level of these emissions. This fear of radioactive contamination was augmented in 1986 by the deposition of radioactive fallout on upland areas of the county following the disastrous explosion at the Chernobyl nuclear power plant in the Ukraine which caused a plume of radioactive material to spread out over large parts of the European land area. As we will see in the first case study in this chapter, the fallout from Chernobyl was complex, both in terms of the physical processes that followed the event and in terms of its economic and political consequences. These complexities meant that judging the risk of the health hazard to humans in the UK was difficult.

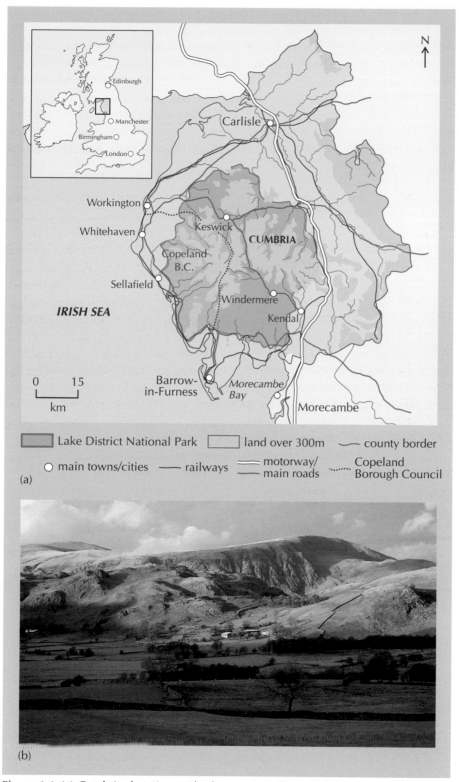

**Figure 1.1** (a) Cumbria: location and relevant sites; (b) typical Cumbrian landscape near Thirlmere.

Further debate and concern over the presence and risks of a radioactive hazard in Cumbria emerged in the 1980s and 1990s when a site for the deep disposal of nuclear waste close to Sellafield was suggested. A debate ensued as to how this site had been chosen, its overall suitability and the means by which the risks could be understood. This case is taken up in the second half of the chapter. It provides a working example of the difficulties in calculating risks over long timescales. Both cases also illustrate the difficulties that are encountered when policy responses are based almost solely upon narrowly defined expert opinion or understanding. They are examples of policy failures from which important lessons can be learned.

In this chapter our aims are to:

1   Understand how environments respond to change in ways that are often complex and dependent upon many processes. The latter range in scale from the atomic to the planetary and include the activities of people, animals, plants and minerals.

2   Understand that this complexity makes it difficult to formulate clear-cut answers as to which course of action will prevent hazards occurring, or at least minimize the risk of their occurrence.

3   Ask questions about the expert-led approaches to formulating policy that are illustrated in the case studies and explore alternative, consensus-led means of response that may lead to more sustainable and fairer outcomes.

While the chapter is concerned with case studies that focus on risks of radioactive-related hazards, you should consider how to apply this understanding to other environmental issues, including debates around the safety of food, climate and various forms of pollution.

# 2    Radioactive lambs: estimating the risks to health in Cumbria after Chernobyl

## 2.1  Understanding radioactivity risks and hazards

On 26 April 1986, Reactor No. 4 in the Chernobyl nuclear power plant in the Ukraine suffered a partial meltdown and the resulting explosion blew a cloud of radioactive dust high into the atmosphere. After travelling 4000 km, circling the European mainland for seven days, under mostly dry conditions, part of the fallout reached the British Isles coinciding with heavy rainfall on its western upland fringes (Wales, Cumbria, parts of Scotland and Northern Ireland): see Figure 1.2. This resulted in large amounts of radioactive material being deposited on the land surface. The question at the time was, what effects would this deposition have? In other words, what were the risks of there being ill-effects from this event? (See Box 1.1 for an explanation of the difference between radioactivity and radiation.)

**Figure 1.2** Map showing the progress of the radioactive plume from Chernobyl, April/May 1986.
*Source:* Gould, 1990, Figure 1.

## Box 1.1  Radioactivity and radiation

*Radioactivity* is caused by the spontaneous disintegration of atoms, and results in the emission of high-energy particles called alpha and beta particles and gamma rays which are similar to x-rays.

*Radiation* is often used to describe these emissions, though strictly speaking the term only applies to electromagnetic waves (gamma and x-rays).

Radioactivity is a feature of most environments, where a background level can be detected using sensitive detection devices, known as Geiger counters, which measure the number of atomic disintegrations per second. Life has developed successfully in the naturally radioactive environment. However, when levels of radioactivity increase above certain levels, the high-energy rays and particles can cause irreversible damage to living tissues.

The first stage in calculating the risk was to measure and map the material that had been deposited. The distribution of radioactivity in England, Wales and Scotland following the event is depicted on the shaded contour map in Figure 1.3 while Box 1.2 explains the ways in which radioactivity and its effects are measured.

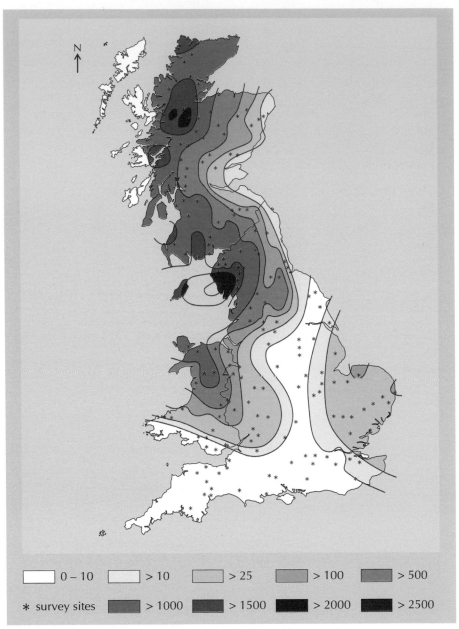

| | | | | | | | | | |
|---|---|---|---|---|---|---|---|---|---|
| 0 – 10 | > 10 | > 25 | > 100 | > 500 |

* survey sites    > 1000    > 1500    > 2000    > 2500

**Figure 1.3** Caesium-137 deposition on vegetation (in becquerels per square metre, Bq $m^{-2}$), May 1986. The map is constructed from measurements of radioactivity taken at a number of survey sites. These measurements form the basis for estimating radioactivity on land in the areas that fall between these sites. An assumption built into the map is that radioactivity will rise or fall evenly between sites of different values.
*Source:* National Environment Research Council, 1998, Figure 1.

## Box 1.2  Measuring radioactivity, absorbed dose and effective dose

*Radioactivity* can be measured by counting the emissions of high-energy particles and rays over a certain area or from a certain mass over a defined time-period. However, this does not necessarily tell us how harmful that radioactivity will be to living organisms. In order to calculate this we need a measure of the *dose* that an organism receives.

*Absorbed dose* estimates how much radioactivity will be received by a living organism. This will depend, amongst other things, on how close the organism is to the source of the radioactivity. However, this measure does not take into account the effectiveness of the dose received. It may make a difference, for example, if radioactive material is eaten rather than absorbed through the skin, and alpha particles cause more damage than beta particles or gamma radiation. *Effective dose* takes some of this complexity into account (see below).

These measurements and their units are explained in some more detail below.

| Unit | Measures |
| --- | --- |
| becquerel (Bq) | *Radioactivity* (the spontaneous decay of atomic nuclei)   This is the measure of the number of atomic disintegrations per second. A becquerel is often related to a measure of area or mass: so for a land surface the number of becquerels per square metre is a normal measure; for food the measure is usually made per kilogramme. |
| gray (Gy) | *Absorbed dose*   The amount of radioactivity absorbed by a body is proportional to its size and inversely proportional to the square of the distance between source and body. Absorbed dose is measured in grays, which are defined as the energy absorbed per unit volume. Dose depends on the nature of the emitter as well as the location and size of the absorber. |
| sievert (Sv) | *Effective dose*   Normally expressed in millisieverts (mSv): $1\text{mSv}=10^{-3}$ or 0.001 Sv. Some radioactive substances can be more damaging to living cells than others. This is in part a function of the relative amounts of alpha and beta particles and gamma radiation emitted. Alpha particles are roughly 20 times more effective at producing damage to living cells than beta particles and gamma rays. Also important is whether the dose is received externally or is inhaled or ingested (eaten) thereby attacking internal organs. A quality factor is introduced to account for these variations. The measurement of effectiveness is also normally expressed over a period of time. The longer the exposure, the more effective it will be. In the UK, the average exposure from background sources is around 2.6mSv per year. These sources are shown in Figure 1.4. People working with radioactive substances are allowed exposure of up to 20 mSv per year. Exposure to doses of a few hundred mSv would cause radiation sickness, |

with a dose of over 1,000 mSv (1 sievert) causing more serious illness and increasing the probability of death: exposure to 4,000 mSv would kill 50 per cent of people.

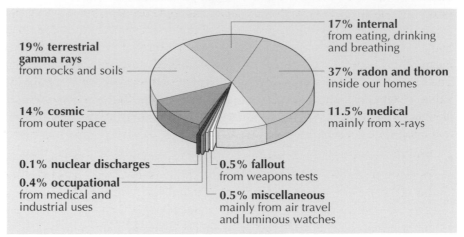

19% **terrestrial gamma rays** from rocks and soils

17% **internal** from eating, drinking and breathing

37% **radon and thoron** inside our homes

14% **cosmic** from outer space

11.5% **medical** mainly from x-rays

0.1% **nuclear discharges**

0.4% **occupational** from medical and industrial uses

0.5% **fallout** from weapons tests

0.5% **miscellaneous** mainly from air travel and luminous watches

**Figure 1.4** Sources of radiation exposure for the general population in the UK: this indicates that only 0.6 per cent comes from nuclear fallout and discharges.

## Activity 1.1

Look at the map in Figure 1.3.

(a) Can you think of any potential limitations in using this map to predict, at any given location and at any given time, the level of radioactivity?

(b) Using the map and Box 1.2, how useful do you think the map is in producing estimates of the risk of contracting an illness caused by exposure to radioactivity?

### Comment

(a) This map is drawn on a large scale which makes it difficult to account for all the variations that can occur when you consider the local circumstances that pertain to a particular site – rainfall can vary over small distances, vegetation varies, as does slope, aspect and so on. It is also a static representation of a situation, giving little indication of how things might change in the future. This map may be useful as a broad guide to the location of fallout, but you should appreciate the difficulty in mapping the spatial and temporal complexity of environmental responses to something like a radioactive fallout event.

(b) The units of measurement – becquerels per square metre ($Bq\,m^{-2}$) – indicate the activity in an area but tell us very little about the kinds of radioactivity present, its possible pathways into living organisms and how effective it will be in causing illness. Therefore in order to achieve a better understanding of the risks to health, more information would be needed.

In order to develop a better picture of the risks from the Chernobyl fallout, a consideration of the particular radioactive substances that were emitted from Chernobyl would be needed. Some substances are more potent than others in that they emit a higher proportion of alpha particles. Some substances decay rapidly, the upside being that they are soon rendered harmless, but the downside being that they may be extremely powerful emitters over their short lifespan. Some substances are concentrated or less easily excreted by living organisms, meaning that they build up in the body and so their effectiveness is compounded. The Chernobyl fallout largely consisted of two **radioactive isotopes**: radioactive iodine (iodine-131) and radioactive caesium (caesium-137). The two isotopes illustrate why the source of the radioactivity matters. There are three **variables** that need to be considered here:

- The *time* taken for the isotope in its unstable form, which is radioactive, to return to its more stable form. This time-period is measured in terms of the **half-life** of the radioactive substance. The half-life is a measure of the time taken for the substance to reduce its radioactivity by half. This is constant for any particular isotope. So, for iodine-131 the half-life is 8 days. This means that if the radioactivity due to iodine-131 is 100 Bq g$^{-1}$ (becquerels per gramme) on day one then eight days later it will be 50 Bq g$^{-1}$, and 16 days later it will be 25 Bq g$^{-1}$. Caesium-137 has a half-life of 30 years, which means that it remains active in the environment for a much longer period (but it emits fewer becquerels per gramme as a result of its lower rate of decay).

- The *type of emission* that a radioactive substance emits. Alpha particles are roughly 20 times more effective at causing cell damage than beta particles or gamma rays. This information can be used to compare the health risks of radioactive substances. Caesium-137, for example, emits beta and gamma rays and so is generally not thought to be as dangerous as some forms of radioactive plutonium, which are alpha emitters. On the other hand, it is also important to note that alpha particles are more easily impeded than the other forms of radioactivity. They can be stopped by a piece of a paper, whereas gamma rays can penetrate all but the most massive of lead-lined concrete structures.

- The *pathway* that the radioactive substance takes in living organisms and in the environment. In terms of pathways in organisms, we can take the example of iodine in people. Ingested iodine accumulates in a particular gland, the thyroid gland, which is located in the neck. Taking iodine tablets, which contain the stable form of the element, can help to flush out and dilute the concentrations of the radioactive substance and the health hazard can be removed from the body relatively quickly. Other radioactive substances are less easily removed and accumulate in organs such as the liver and skin, and may therefore remain effective in the body for longer periods.

Pathways are equally important when considering environments. For example, some radioactive substances form soluble salts in water and may

---

**Isotope** is a term used to refer to different forms of the same element. More specifically it refers to atoms with the same atomic number but different numbers of neutrons and therefore a different atomic mass. **Radioactive isotope** is a term used to refer to an unstable form of an isotope, the form being indicated by a number which refers to the atomic mass.

**Variables** are aspects of a situation that can vary in terms of properties, intensity or other measurable characteristics.

half-life

Bioaccumulation is discussed in **Drake and Freeland (2003)**.

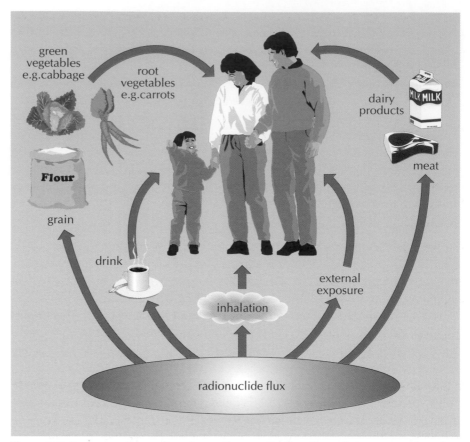

**Figure 1.5** Some possible radioactive pathways. Radioactive substances can be taken in through consuming food or water contaminated by exposure to radioactivity; humans may also be exposed to external sources of radioactivity directly or through breathing in microscopic particles.
*Source:* Nirex, 1995, Figure 1.

therefore be easily diluted and dispersed. Caesium-137 is one such substance. The Sellafield pipeline, which emits wastes into the Irish Sea, discharges this isotope of caesium (along with a number of other radioactive substances): it is widely thought that, since it reacts with water to form a soluble compound, it is quickly dispersed and diluted to non-hazardous concentrations by marine processes. Conversely, many plutonium isotopes are insoluble and it is now recognized that marine tides and currents can concentrate these substances in nearby estuaries, including Morecambe Bay (see Figure 1.1a). Here they are available to living organisms, especially bottom feeders such as shellfish, and their uptake and storage in living tissues can result in even higher concentrations. The pathway does not stop there but goes up the food chain to the human population. Some are more at risk than others. For example, those people who spend time on the estuary and those who eat the local shellfish are more prone to exposure to the radioactivity and radiation. Pathways can therefore be affected by social circumstances and habits as well as chemical, physical and biological

processes. These at-risk groups are often termed **critical groups** : they are     critical groups
thought to be at greatest risk of suffering ill health. If it can be demonstrated
that the critical group is not at risk, then it is usually accepted that the risk to
the public at large will be minimal.

### Activity 1.2

Which of the following would you say is likely to be the most difficult to
measure or predict with any certainty:

- the half-life of a substance
- the type of particles and waves emitted by that substance, or
- the pathways that it can take in environments and in living organisms?

### Comment

The half-life and the chemical make-up are both likely to be known, and can
be used to estimate risks in a fairly straightforward manner. It is the pathways
that are likely to be more complex and uncertain. Compared to the relatively
predictable nature of radioactive decay, pathways can be more open-ended,
illustrating the multitude of physical, chemical and biological as well as social
processes that are involved in environmental responses.

In this subsection we have built up an understanding of the complexities which
apply to understanding the degree of risk that radioactive substances can pose in
an environment. Hazard and risk are dependent on the half-life, the type of
emissions and the pathways that are taken by the substances. We have started to
touch on some of the issues that make predicting pathways difficult. The physical
and social complexity of pathways is one element in answering our first aim which
was to understand how environmental responses involve all manner of people,
institutions, physical processes, organisms, biological processes and so on. In the
next subsection we will be building on this by investigating some of the difficulties
encountered in measuring and predicting the environmental responses to the
Chernobyl event in Cumbria. We will also be touching on our second aim, which
is to explore the difficulties involved in responding to these complex
environmental hazards.

## 2.2 (Mis)understanding the risks

The deposition of fallout over the Cumbrian hills can be termed a potentially
hazardous event. Unlike the deposition of radioactive substances in the
immediate surroundings of the Chernobyl plant, the degree of hazard and the
risks of the hazard producing ill-effects were not immediately apparent. At
Chernobyl, 45 people received doses above 4,000 mSv (see Box 1.2) in a short
period of time, within months 31 had died, and the still high levels of radioactive
substances continue to contribute to cancers and birth abnormalities. The risks
were high and the hazard was well known (although the Soviet government was

initially slow to reveal the extent of the disaster to the population and to the international community).

In Cumbria things were less certain. The doses to people were much lower, estimated at 0.27 mSv per person living in the UK over the year following the event (which is roughly 10 per cent of the annual average dose from background sources). The main concern was the presence of caesium-137 with its relatively long half-life. The worry was that grazing animals, and particularly sheep, would eat contaminated vegetation and the radioactive substances would be incorporated into their tissues and so enter the human food chain. Two properties of caesium-137 were used to estimate the likely risks of this occurring. First, its solubility in water led to expectations that the heavy rains which deposited the fallout would also wash much of the radioactive material off the land surface into ground water where it would be unavailable to grazing animals. Second, any remaining caesium-137 was expected to adsorb onto clay particles in the soil. As **Morris (2003)** demonstrates, adsorption is an electrostatic bond that holds the positively charged cations, in this case the caesium, in the soil. These cations are available for plant uptake, via cation exchange, although the degree of exchange was expected to be low given the strength of the bonds. Therefore, if plant uptake of caesium-137 was low, then the amount reaching the grazing animals would also be low. (We will see later on that this assumption of low uptake by plants was particularly questionable in the Cumbria situation.)

See **Morris (2003)** for a discussion of soil processes.

There were also economic issues that could be incorporated into an estimation of risks. The sheep and lambs grazing in this area would not normally go to market until late in the summer, by which time it was expected that the radioactivity would have diminished in the sheep meat and in the local environment. The risk of the radioactivity reaching the human food chain would be – initially at least – very low. Given these considerations, it was not at all obvious that there was a problem.

With these kinds of issues in mind, government ministers and Ministry of Agriculture scientists assured the general public and farmers that there was no need for concern. The hazard, if there was one, would soon be over and the risks of ill health would become very small indeed. However, within a few weeks some of these assurances started to look questionable. Radioactivity levels started to increase in hill sheep. Ministry scientists continued to tell farmers that levels would soon show a decrease. But radioactivity levels kept rising, so that by 20 June 1986 (three weeks after the fallout), the Secretary of State for Agriculture saw fit to impose a ban on the movement of sheep in Cumbria. The ban was initially for three weeks, and farmers were told that no financial compensation was necessary because it was only a temporary issue. As it turned out, even fifteen years later, levels of radioactivity in Cumbrian sheep were still reaching unsafe levels.

The environment was not responding according to plan, or according to the understanding or knowledge that had led Ministers and government scientists to assume that the problem was short-term. With the benefit of hindsight, this may not be altogether surprising. Environments are complex and open to all manner of processes of which many are understood but many others are unknown. In addition, as we have already noted, the pathways that radioactivity can take are often highly complex and subject to uncertainties (inaccuracies, ignorance and indeterminacies: see Box 1.3).

## Box 1.3  Uncertainties

Uncertainties can include the following:

**Inaccuracies** – where measurements are subject to errors.                    inaccuracies

**Ignorance** – or the not yet known, where we are waiting on new               ignorance
information or experimental results. These are gaps in our understanding,
waiting to be filled.

**Indeterminacy** – or the unknowable. This is particularly important when      indeterminacy
dealing with issues of such complexity or novelty that it is impossible to
know for certain what the outcome will be.

The fact that the environment responded differently to expectations may not at first sight give us too much cause for concern. After all, the situation regarding the radioactivity of sheep meat was monitored and when levels exceeded safe limits a ban was put in place. No-one was knowingly exposed to unsafe food. However, through interviews and focus groups, both of which sought to engage people in discussions of their concerns and views of events, researchers have raised questions over the effectiveness of the policy responses to the Chernobyl fallout (see Wynne, 1996a). This research is not a conventional survey using large numbers of respondents, but instead acquires its significance from analyses of the ways in which respondents talk about their experiences, and how these relate to others' experiences and to broader social and cultural processes.

The research highlighted the various controversies that followed the Chernobyl event. Farmers, for example, were placed under a good deal of stress by reassurances followed by policy reversals. Some lost out on compensation because they were repeatedly told that the ban was temporary and that therefore they should hold on to their lambs for a few weeks more. When it became evident that the ban was more permanent, their lambs had passed their prime and therefore their compensation value had decreased. Farmers also reported frustration and suggested that officials were arrogant in that no-one had deigned to ask for their views on the situation. They accused government scientists of complicity with the Ministry officials whom they assumed wanted to minimize compensation claims. They even started to suggest that radioactivity levels may well have been high for some time before the Chernobyl event and were in fact related to the Sellafield site. They suspected that blaming the radioactivity on Chernobyl fallout was a convenient cover-up for a long-standing pollution hazard. In short, there was a breakdown in trust between farmers and government advisors. Farmers would view almost anything suggested by representatives of

**Figure 1.6** Cumbrian farmers began to suspect that high radioactivity levels in sheep were related to the nuclear reprocessing plant at Sellafield.

the Ministry of Agriculture with a good deal of suspicion. Future working relations between government representatives and farmers were jeopardized.

In view of these concerns, the responses of policy-makers to the post-Chernobyl fallout in Cumbria have now become matters for careful analysis and discussion. A number of sociologists have argued that avoidable mistakes were made and these can be analysed in order to point towards more effective ways of responding to hazardous events where the extent of the hazard and the risks of harm are subject to uncertainties (see Wynne, 1996a, b and Irwin, 1995). Two issues are of particular concern. The first is the relationship between scientists, scientific understanding and policy, and the second is the relationship between the lay public and scientific and governmental experts. Each speaks to our aims of understanding the response to environmental changes and exploring alternative means of making policy.

## Scientists, science and policy responses

Earlier, it was noted that the official understanding of the low level of risk following the fallout rested on a number of assumptions. The most problematic of these turned out to be the pathway that the caesium-137 would take in the soils. It was shown earlier that pathways are especially difficult to predict. Yet the uncertainties surrounding pathway prediction were not made clear when the policy-makers issued their advice to farmers that there was no risk to their sheep from the fallout (or to their economic value). Sociologists have argued that, in retrospect, it would have been more beneficial to inform farmers and the public that, whilst the risks were thought to be low, government scientists could not be sure and therefore a precautionary approach was being taken. This approach would have involved placing an immediate ban on the movement of sheep until such time as there was evidence that there was no risk.

See **Bingham and Blackmore (2003)** on precautionary approaches to policy-making.

This is not an isolated case. An under-reporting of uncertainties has marked many UK governmental responses to environmental hazards. In 2000 the UK government was criticized by the members of the BSE (Bovine Spongiform Encephalopathy) Inquiry for ignoring uncertainties when reassuring the public that British beef was safe to eat in the 1980s and 1990s. Likewise, anti-GMO campaigners have criticized governments and industry for downplaying the uncertainties about the environmental consequences of farm trials and commercial-scale planting of genetically modified crops (see **Bingham, 2003**). This tendency to ignore uncertainties caused so much concern that in 1997 the then UK government's Chief Scientific Advisor, Robert May, drafted a paper entitled 'The use of scientific advice in policy making' (May, 1997). In the paper May called for a more mature relationship between scientists and policy-makers. He argued for a rejection of the practice whereby carefully selected scientists were pressured to supply unequivocal advice or facts which policy-makers could then use on a one-off basis to define policy. In its place he argued for policy-makers to seek a wide body of sometimes conflicting scientific opinion and for both scientists and policy-makers to state the uncertainties that they recognized or could envisage. May recommended that scientists should become more involved in the design, implementation and continual reassessment of policy.

May's identification of a problem in the relationship between science, scientists and policy is evident in our case of the fallout event. In assessing the pathways and properties of caesium-137 policy-makers and government scientists drew upon one set of scientific experiments carried out on an airfield, at Harwell in Oxfordshire in the 1960s. The policy-makers neglected to investigate two sets of uncertainties: first the uncertainties (in the form of inaccuracies and ignorance) reported by Harwell researchers relating to the robustness of their findings and the possible limitations of their knowledge; second, the uncertainties (in the sense of indeterminacies) that accompanied the translation of this carefully controlled experimental knowledge to the more open conditions of post-Chernobyl Cumbria.

To be sure, the Harwell experiments were well regarded. Columns of various soil types were treated at the surface with caesium-137 and then monitored over a period of nearly five years. Grass was grown on each plot and the soils were free-draining. The main findings were, first, that radioactivity at or near the soil surface decreased rapidly over time, for all the soils tested, and, second, that the radioactive substance moved down the soil in rainwater to a mean depth that was roughly the same for all of the soils investigated.

Reread Box 1.2 on absorbed dose. In terms of the aims of the Harwell research, what do you think the mean depth result allowed people to say about the risk of ill-health to people (or any other living organism at the land surface) for the different soil types?

### Comment

If the mean depth of radioactive substances was the same, then the absorbed dose, which is inversely proportional to the square of the distance between the living organism and the radioactive source, will be the same. The risk of ill-health would therefore be the same for all the soils.

The fact that levels of caesium-137 at the surface decreased rapidly *suggested* that any risk of ill health following a fallout event would be short-lived. The similar mean depths of radioactivity in all the experimental soils *suggested* that caesium-137 behaved similarly in all the soils. The lack of radioactivity detected at the land surface throughout the study period supported the assumption that caesium-137 was quickly adsorbed by soil particles and was not normally available to plants through the process of cation exchange.

acid peaty soil    clay-rich soil

- sample depths with significant radioactivity
---- mean depth of caesium-137

**Figure 1.7** Schematic diagram to demonstrate how the range of depths can differ widely whilst the mean depth stays the same for two soil profiles.
*Source:* Wynne, 1992, p.122.

However, there were uncertainties expressed by the researchers. Two are important for our purposes. First, the researchers reported doubts over the accuracy of their measurement in the experimental peat soil, which had a tendency to crumble when being handled. This meant that the mean depth of the radioactivity was difficult to measure consistently, 'resulting in uncertainty' (Gale et al., 1964, p.258). Second, they noted that the range of depths at which radioactivity was recorded was much greater for the peat soil than for the other soils. This increased range did not affect the *mean* depth, which was broadly similar for all soils: see Figure 1.7.

The researchers reported that despite this difference, the mean value was *probably* valid (i.e. this was an educated guess but they were rightly drawing readers' attention to some degree of uncertainty in their knowledge) (Gale et al., 1964; see also Wynne, 1992).

### Activity 1.4

For the two examples of uncertainty discussed above, use Box 1.3 to specify the kinds of uncertainty that the researchers were reporting.

### Comment

The problem of measuring radioactivity in peat is an example of *inaccuracy*. The greater range of depths of radioactivity in peat suggests either further *inaccuracy*, or that something else is going on in this soil that the researchers are not sure about. The latter may indicate an unknown aspect of caesium-137 behaviour and constitutes *ignorance*.

As we mentioned above, these uncertainties were unidentified, ignored or regarded as irrelevant by those people who were making policy following the fallout event in Cumbria. In addition, the added uncertainties, which come into play when experimental results are translated from one environment (the airfield at Harwell) to another (upland fells), were ignored.

One way of making the latter more explicit is to note the major differences between the experimental situation and the situation to which this knowledge is being applied. We can attempt to do this for the Cumbria case by looking more closely at the upland environments.

### Activity 1.5

Study the photograph in Figure 1.8(a), the soil map in Figure 1.8(b) and the 1:25,000 Ordnance Survey map in Figure 1.8(c).

(a) Using the information in front of you, produce a summary of some of the environmental conditions and components in this upland area of Cumbria under the following headings: vegetation, soils, drainage, and topography. These can be quite brief, two or three word summaries (e.g. for vegetation you might write 'mainly trees' or 'all grass').

(b) Using these brief notes, compile a list of those environmental conditions or components that might be important when trying to apply the results of the Harwell experiments to the fell environment.

### Comment

While it is difficult to tell from a single photograph, the vegetation comprises a mix of grasses, sedges (Cyperaceae) and heathers (Ericaceae *Calluna vulgaris*). The soils are largely peats or peat-topped. The surface drainage is via several small streams which are widely spaced. There are many areas with no obvious surface drainage. The topography is varied comprising hills, hollows and gulleys.

The following aspects of the fell environment might make the translation of the Harwell results problematic:

- The soils are largely peat-based. The researchers at Harwell were least certain when it came to the behaviour of caesium-137 in peats.
- The land is unevenly drained. Contaminated water might therefore stand for long periods at the surface.
- The land surface is uneven (and very unlike the airfield at Harwell). Water is likely to run into hollows, where radioactive substances might be concentrated.

- The vegetation includes plants other than grasses, notably heather. The effect that different vegetation cover might have was not considered at Harwell.

You may have thought of other differences. The main conclusion is that the environment in Cumbria is more varied and open to more influences (like different species of plant and animal) than could possibly be the case at Harwell. This is not to dismiss the science: it is normal to be selective when investigating environmental processes. Rather it is incumbent on scientists and policy-makers to acknowledge the necessary selectivity of their knowledge.

**Figure 1.8** (a) An example of topography and vegetation of Cumbrian hills.

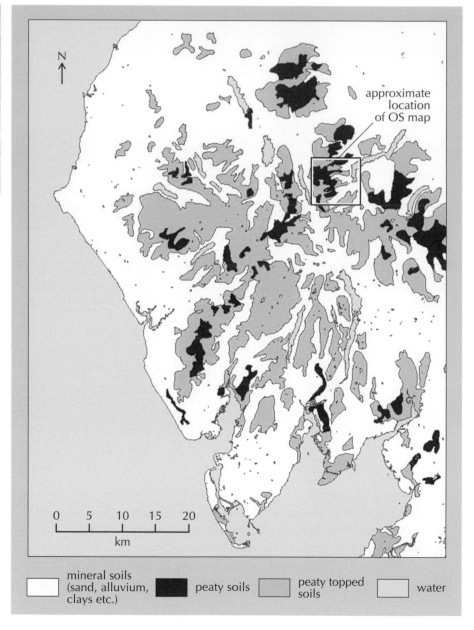

**Figure 1.8** (b) A soil map of Cumbria (© copyright Cranfield University 2003).

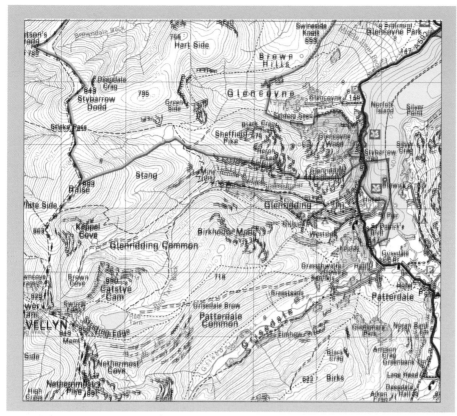

**Figure 1.8** (c) Ordnance Survey map of the Birkhouse Moor area in Cumbria.

Since Chernobyl, research has indeed established that upland peaty soils are different from lowland clay-rich soils in terms of the behaviour of caesium-137. This is largely a result of the amount of clay in the soil. As noted earlier, clay minerals can adsorb caesium-137 onto soil particles and, in doing so, make it relatively unavailable to plants. However, in soils that have low quantities of clay and/or are dominated by organic materials, caesium-137 is not readily adsorbed. Meanwhile, if there is poor drainage, water containing available isotopes remains in the soil for long periods, and is therefore incorporated into new plant growth. All of these circumstances applied to the upland peats. In addition, some vegetation is more effective at taking up caesium from the soil. Heather, it turns out, is particularly effective at taking up caesium-137 through its roots. Sheep and other grazers (including heather-eaters such as grouse) can in turn ingest the radioactive substances and incorporate them into their tissues.

In short, the upland fells were both sufficiently different to the experimental environment and sufficiently varied in themselves to make the prediction of the behaviour of caesium-137 highly problematic. These differences and complexities made any risk assessment an uncertain business. For the Chief Scientific Advisor, there would be more chance of these uncertainties making it onto the policy agenda if a varied body of scientists was involved in the making of policy and if the uncertainties, in the form of inaccuracies, ignorance and indeterminacies, were stated upfront. In the next subsection we will consider another – complementary – suggestion, that of incorporating non-science experts into a policy forum so that a greater range of uncertainties might be aired and to avoid the the kind of collapse in trust and working relations that followed the fallout in Cumbria.

## Lay–expert relations

May's prescription for better policy-making is to include more scientific voices, more often, and to incorporate explicit consideration of uncertainties. A number of other commentators would go further and suggest that we not only need more scientists to guide policy, we also need other professionals and members of the public to consider how risks are estimated and how policies are being designed. Again, the radioactive fallout event in Cumbria has been used to make the case.

○    Review your list of environmental conditions in the previous activity. Who would you say were the experts in these environments?

●    Local land-users, including farmers, would probably be thought of as experts in the local environment.

After the fallout, the policy relationship was *largely* one of advice being passed down from government ministries to people living in the county. We can characterize this as a **top-down** approach to environmental policy responses. It is also sometimes described as following a **deficit approach** with regard to the way in which it treats the public. A deficit approach assumes that the public is ignorant in scientific and environmental matters (see Irwin, 2001, p.95). It assumes that the public needs to be educated and persuaded of the right course of action, and/or that policy-makers already know what people are concerned about and what it is that they want. Because the information largely flows in one direction (from top to bottom), it is often assumed to be the most cost and time effective form of policy-making.

The example of sheep-farming after Chernobyl suggests, however, that the top-down approach, whilst being cheap and efficient in terms of numbers of people involved, can turn out to be an inefficient means of responding to potential environmental hazards and estimating risks. We have already seen how official policy-makers underestimated the importance of local environmental conditions in their policy-making. This resulted in a long and painful delay in developing the right policy. Equally costly were the top-down approaches to management of sheep in affected areas. A few months after the fallout had reached Cumbria, farmers were told to move their sheep to lower-altitude pastures where contamination was lower. However, this short-term fix was rejected by farmers who needed to think about future stock issues. Moving upland sheep to lowland fields would deplete the pasture in the valleys, which was needed for winter hay and silage. Official experts acknowledged this and advised farmers to import feed straw. As one farmer put it when interviewed by researchers:

top-down

deficit approach

[The experts] don't understand our way of life. They think you stand at the fell bottom and wave a handkerchief and all the sheep come running ... I've never heard of a sheep that would even look at straw as fodder. When you hear things like that it makes your hair stand on end. You just wonder, what the hell are these blokes talking about?

(Wynne, 1990, p.150)

**Figure 1.9** Hill-farmer in Newlands Valley. Sheep-farming is an expert practice which can have a significant input to policy debates.

The collapse in working relations suggested that this form of policy response was inappropriate to the task at hand. An alternative would be to turn the tables and pursue a **bottom-up** approach to policy. This would involve allowing those affected by the decisions, like sheep-farmers, much greater involvement in policy-making.

bottom-up

Following a bottom-up model, farmers would have brought important knowledge to the policy discussions. Their knowledge of local environments, of sheep movements and of farm management practices would all have helped in the estimation of risks and environmental uncertainties and in the design of effective policy. However, the problems with this should be immediately apparent. Sheep-farmers may not be in a position to assess risks and their uncertainties in the ways that other members of the public would want. They are experts in raising sheep but not necessarily experts in plotting and understanding the pathways taken by radioactive substances. Likewise, sheep farmers may be perceived as potentially embodying a clash of interests between making their businesses profitable and ensuring the safety of consumers.

See **Burgess (2003)** for examples of ways in which deliberative forms of decision-making can be used to value environments.

There are a number of suggestions for avoiding the pitfalls of both top-down and bottom-up approaches. Most include an attempt to incorporate a range of views from various **stakeholders** (those with a specific interest in an issue), who can comment on each others' proposals and policy recommendations. Many of these

stakeholder

are termed *deliberative* forms of decision-making. Deliberative democracy will be discussed later in this chapter. For the moment it is worth simply noting that it refers to attempts to develop more open forms of policy-making and decision-making wherein groups of people present and discuss various perspectives on an issue and then attempt to move to a consensus or compromise solution. The idea is to incorporate as many relevant perspectives as possible without assuming at the outset that one group or one kind of approach is the only valid approach (which is not to suggest that all views are equally valid).

Whilst by no means perfect, deliberative forms of decision-making can help groups to bring concerns to the decision-making forum that might otherwise be neglected. At their best they can lead to frank discussions and to a fuller consideration of the uncertainties that might exist. In this way, proponents argue, social responses to potential environmental hazards can be more sensitive to, and responsive towards, the complex ways in which environments respond to change. At their worst they can be extremely slow and expensive ways of making decisions and sometimes may not in the end manage to broker an agreement between conflicting parties.

## Summary

In this section we have used a detailed discussion of a specific issue to illustrate the kinds of approaches that are used to understand the risks associated with potentially hazardous environmental changes. We have demonstrated that the environmental responses to the fallout event were physically and socially complex – so much so that the pollution pathways were difficult to determine with any certainty. This relates to our first and second aims which were set out in the introduction. We have also started to demonstrate that finding responsive ways of dealing with these uncertainties has led some to call for a better relationship between science and policy-makers, and others have called for a broadening of what we call 'expertise'. Both amount to a redefinition of the policy process as more open and deliberative. We have hinted at some of the difficulties of such proposals.

The next section introduces a second case study which takes these questions forward, and in so doing furthers our third aim which is to explore alternative ways of responding to environmental issues.

## 3  The politics of response: why plans for deep disposal failed at Sellafield

Uncertainty is a common characteristic of scientific inquiry, affecting all our attempts to understand environmental processes and change. Yet we have argued that conventional approaches to policy may neglect, ignore or be unaware of variables or conditions that are recognized and understood by both scientists and those lay people who possess a routine understanding of risks and hazards derived from their experience. This has led us to suggest approaches to risk estimations which combine both expert and relevant lay knowledge. Such an endeavour is likely to yield benefits in two respects. First, it will add to public understanding of risk and hazard and ensure greater credibility for scientific findings and predictions. And, second, by embracing a broader range of knowledge and experience, it will help to develop more responsive policies and achieve political legitimation.

These issues are further explored here in another nuclear case study also located in Cumbria. It is the case of the Rock Characterisation Facility (RCF) or underground laboratory that was proposed for investigating the geological conditions for the deep disposal of solid radioactive wastes. The bulk of radioactive wastes arise from the production of nuclear energy during the lifetime of a nuclear reactor (approximately 40 years) and from the **reprocessing** operations which are used in the UK to recover plutonium and uranium. Wastes also come from military sources especially nuclear submarines and are produced from the use of radioactive materials for medical and research purposes. Radioactive wastes present various levels of hazard so must be carefully managed. The safe management of these wastes, minimizing the risks that they may pose to present and future generations, is the topic of this section.

**Reprocessing** is the chemical processing or recycling of spent nuclear fuel to recover uranium and plutonium.

See **Blowers and Elliott (2003)** for a discussion of radioactive wastes, including the classification of wastes and for estimates of the volumes of each type of waste.

## 3.1 Future hazards

So far in this chapter we have been talking about hazards to the present human population, but some of the radioactive substances in nuclear wastes have half-lives that far exceed the 30 years for caesium-137: plutonium-239 has a half-life of 24,000 years, iodine-129 has one of 15.7 Ma (Ma = million years) and uranium-238's half-life is 4,500 Ma. Such time horizons pose profound challenges to planning safe management of nuclear wastes. As you will appreciate, the dynamic nature of environments means that whether or not these wastes are going to be safe in the future is a matter that is surrounded by considerable uncertainty. (We shall look again at the implications of these timescales in section 3.4 below.)

Dealing with these wastes over such long periods of time in ways that minimize the risks of hazard is a scientific, technical and social issue. The RCF represents an attempt to find a solution which, like the sheep-farming case, demonstrates that policy failure resulted from poor procedures, a tendency to downplay uncertainties and an underestimation of the importance of expert–public dialogue.

The RCF was intended to provide geological data that could be used to determine the feasibility of developing a 700m deep repository that would enable solid radioactive wastes to be isolated and contained. The wastes in question were intermediate and low level wastes (ILW and LLW) which were accumulating in temporary stores at Sellafield (high level wastes – HLW – were to be left in surface stores for at least fifty years). The RCF was proposed as a means of reducing some of the uncertainties surrounding the estimation of a future risk of radioactive contamination from these wastes.

We will review this case in three steps. First, we will look at the ways in which the Sellafield repository site was chosen. Second, we will look at the problems at the site and ask why it failed. Finally, we will look for lessons from this process. The three parts map onto our three aims of looking at the social and physical complexity of environmental responses, looking at the difficulties that this complexity produces and asking questions as to what kinds of policy responses might be preferred in the future.

## 3.2  Narrowing the options: Sellafield – the preferred or rational choice?

Nirex was formerly NIREX, the Nuclear Industry Radioactive Waste Executive but was renamed and put on a commercial footing in 1985.

During the 1980s Nirex had set out to find and develop suitable disposal sites for intermediate and low level wastes, but had encountered opposition to its plans for repositories in Eastern England (see **Blowers and Smith (2003)** and **Blowers and Elliott (2003)**). Nirex had failed to achieve its objectives for a number of reasons. The company had operated an approach that has been termed Decide–Announce–Defend or DAD which, essentially, involves picking sites in secret and then 'consulting' (after the event, so to speak) with the communities in the selected areas. This approach meant that local groups were immediately placed on a defensive footing. They mobilized resources and built coalitions that influenced the government to put pressure on Nirex to withdraw the proposals.

By 1987 it was decided to begin afresh the search for a site for the deep disposal of the wastes. Nirex published *The Way Forward*, a consultation document, to which responses were requested, signalling its intention 'to promote public understanding of the issues involved and to stimulate comment which will assist Nirex in developing acceptable proposals' (Nirex, 1987, p.4).

### Activity 1.7

Referring to the earlier discussion of lay–expert relationships, how would you characterize first, Nirex's DAD approach and, second, the approach suggested in *The Way Forward*?

### Comment

The DAD approach is an example of top-down policy-making. The approach in *The Way Forward* is, to some extent at least, more deliberative in intent, although the phrase 'promote public understanding of the issues' still sounds as if elements of a deficit model remain.

In *The Way Forward* Nirex indicated that there were five types of geological environment likely to offer potential sites: see Figure 1.10. A suitable environment was likely to be where there were certain rock or geological formations that could inhibit the flow of groundwater and which would therefore reduce the risks of radioactive substances being conveyed to the Earth's surface. These five types of suitable environment together covered around 30 per cent of the deep geology underlying the land area of the country, in which 500 possible sites were identified. To narrow the field, areas of national conservation importance and those with population densities above 5 persons per hectare were eliminated. This reduced the number to 200 and these were progressively refined until there were 39 sites. Through further elimination the desk studies identified a shortlist of 12 sites, 10 on land and two offshore. These twelve sites were then subjected to a multi-attribute decision analysis (MADA) (see Box 1.4). From this 'a picture emerged of three sites that could be recommended for further investigation, with possible merit in a further two' (Phillips, 1995, p.7). Among these were Dounreay, in Scotland, and Sellafield.

## Box 1.4   Multi-attribute decision analysis (MADA)

The MADA procedure is in practice extremely complex and so can only be briefly summarized here. In the case of the Nirex siting exercise it involved a dozen participants drawn from Nirex and from experts in geology, transport and planning. Thirty attributes or criteria relating to repository design and siting were identified and organized into four main groups covering costs, robustness, safety and environment. The participants, working in groups, assigned weights to the attributes which were compared and eventually the relative importance of each attribute was agreed by the whole group to create a base model. The attributes were then applied to the sites using different weighting systems.

In addition to this expert-led decision-making procedure, Nirex distributed copies of *The Way Forward* to national and local authorities, professional bodies, environmental groups and the public. The company also held over sixty seminars and meetings and received over 2500 written responses which were analysed by the University of East Anglia to elicit views on a range of disposal concepts and potential areas of search for sites. There was no unanimity of view though there was little support for offshore disposal. On the basis of the MADA exercise and the consultation, Nirex decided to limit its investigations to 'two areas where there is a measure of public support', namely Dounreay and Sellafield (Nirex, 1989, p.51), and in 1989 recommended them as sites for investigation for the deep disposal of ILW and LLW. Subsequently, in 1991, Dounreay was dropped on grounds of the much higher transport costs to such a remote location, leaving deep disposal at Sellafield as the preferred – indeed the only declared– option for the future management of the UK's ILW and LLW.

74725

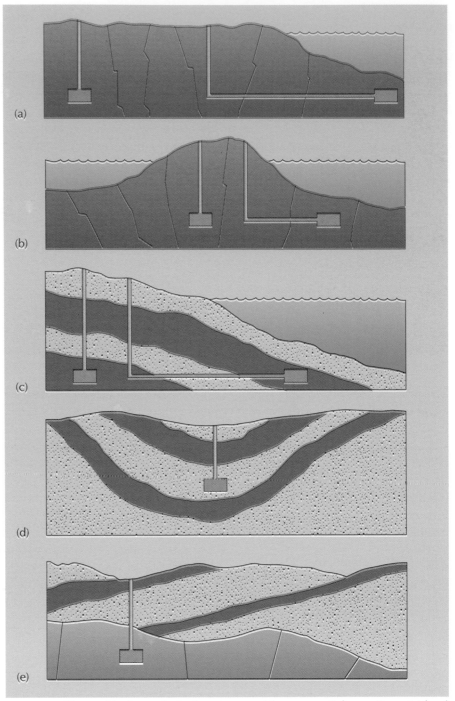

**Figure 1.10** Types of hydrogeological environment for a potential repository: (a) hard rocks in low relief terrain; (b) small islands; (c) seaward dipping and offshore sediments; (d) inland basin; (e) basement [rocks] under sedimentary cover (BUSC). Each theoretically results in a low risk of contaminated groundwater reaching the surface. The 12 sites selected contained examples of each of these types of environment. *Source:* Nirex (no date), p.17.

**Figure 1.11** Sellafield reprocessing works.

The process above may sound logical and fair. However, there is evidence to suggest that the choice of Sellafield depended on the way in which the selection criteria were manipulated. Indeed, the specific site eventually chosen for the RCF – at Longlands Farm near Sellafield – was not initially included and was actually added in, it would seem, when the list was reduced to 39. It was introduced as offering better geological potential in some respects than a nearby Sellafield site originally identified. Having made a late appearance, Longlands Farm survived further sifting as a site that, while not necessarily the best, would be expected comfortably to meet the necessary safety standards. Since all the sites that subsequently went through the MADA exercise were expected to meet the long-term safety target, this would not prove a discriminatory variable. Conversely, cost and safety factors (both conventional and radiological) associated with waste transport were given a high weighting and this especially favoured Sellafield, where most of the wastes were already located. The weightings applied depended to a large degree on the judgements made by the participants in the MADA exercise. The MADA and the consultation prepared the ground for what was essentially a pragmatic decision reflecting political expediency. The point was made by the then Secretary of State for the Environment, Nicholas Ridley, that 'it would be best to explore first those sites where there is some measure of local support for civil nuclear activities'. Political convenience and economic costs may not in themselves be poor criteria for selecting sites. However, people's unease centred on the suspicion that the scientific criteria for safe disposal had been somewhat compromised by these commercial and political concerns.

Despite the political expediency of the decision, the proposal met with mixed reactions. On the one hand, opinion polls indicated considerable support. A survey in 1992 for Copeland Borough Council (the local authority covering Sellafield) indicated half in favour of the repository and a third opposed from a sample which included 42 per cent of respondents dependent on the nuclear industry. A Gallup poll taken for Nirex in 1995 showed even higher support in Copeland (69 per cent for, 19 per cent against) and still over half in favour in the wider area of Cumbria. On the other hand, these statistics masked much more complex attitudes in a local community that combined a traditional farming and fishing population with those working in the industry, as well as a growing population entering the area to work in tourist services and to retire or otherwise enjoy the regional amenities. Within this mixed community, complex feelings were revealed in research using qualitative interviews. People were likely to be fatalistic (believing it will come here anyway), ambivalent and acquiescent. There was a belief that the project was destined to end up in Cumbria because it had been rejected elsewhere and 'the local population was too subservient and browbeaten to do likewise' (Wynne et al., 1993).

While local support for the Nirex project was qualified, those living farther away, less dependent and less involved in Sellafield and the nuclear industry, expressed strong and determined opposition. More importantly, from a political point of view, the opposition included Cumbria County Council and local and national environmental groups.

Having obtained planning permission for exploratory drilling, Nirex began an extensive survey of the geology of the area. In line with international practice, it was soon declared that the long-term safety of a repository at the site would be difficult to test in the absence of detailed geological information and that this could only be obtained through constructing an *in situ* laboratory. Accordingly, in 1992 the company applied for permission for a Rock Characterisation Facility (see Figures 1.13 a and b) but this was turned down by Cumbria County Council. Nirex appealed against the decision and a local public inquiry lasting 66 days spanning 1995/6 was held to determine the matter (Figure 1.12). The Inspector who presided over the Inquiry recommended against the proposal and this advice was upheld by the Secretary of State for the Environment, John Gummer, who dismissed the appeal on the eve of the General Election in March 1997.

**Figure 1.12** Protesters at the start of the public inquiry into the Rock Characterisation Facility proposal.
*Source: New Scientist,* 6 January 1996.

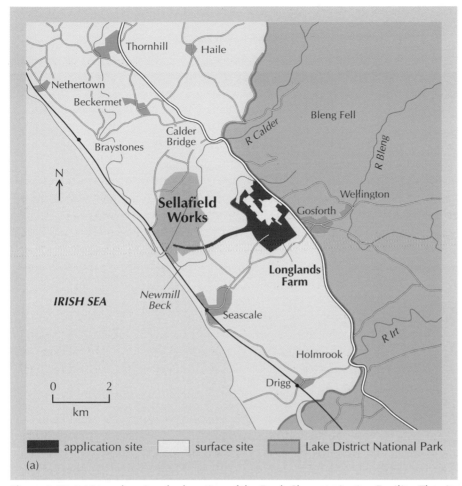

(a)

**Figure 1.13** (a) Map showing the location of the Rock Characterisation Facility. The site at Longlands Farm on land owned by BNFL is just east of Sellafield and on the border of the National Park.

So far in this section we have examined how Sellafield came to be chosen. Following the defeat of its DAD strategy in the 1980s Nirex had embarked on a more consultative approach in conjunction with a process for site selection using a variety of criteria. But, despite *The Way Forward*, the company, in 1997, was no further forward in finding an acceptable way of managing radioactive wastes in the long term. In the next subsection we explore why Nirex failed to make the case for the RCF at Sellafield and consider the problem of uncertainty in decision-making involving risk over long timescales. This involves us raising questions, in line with our third aim, concerning conventional forms of policy-making, and looking at the approaches that might be devised to find better ways of responding to long-term risks.

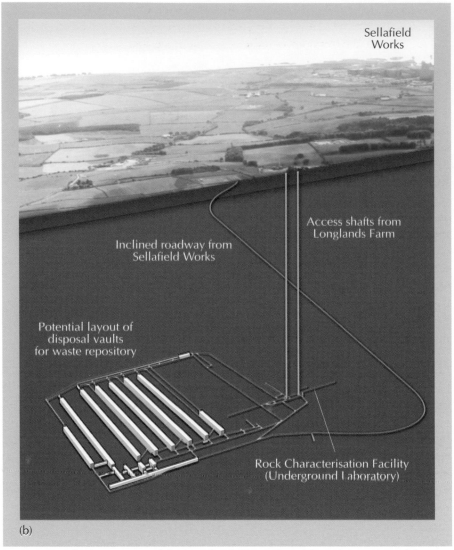

**Figure 1.13** (b) A cut-away of the Longlands Farm site with an impression of the two deep shafts and underground RCF laboratory, as well as the proposed repository and access, at approximately 735 metres depth.
*Source:* (a) Nirex, 1994; (b) Nirex (supplied to the authors), 2003.

## 3.3  Questioning scientific uncertainties and selection processes

In dismissing the appeal on the RCF in 1997 the Secretary of State commented on the poor quality of the proposal in terms of conventional criteria such as the visual impact, the poor local access and adverse effects on the local environment. However, the case for rejection went much further. The Minister had two further concerns. One was about 'the scientific uncertainties and technical deficiencies in the proposals presented by Nirex'. The other was a concern 'about the process of the selection of the site and the broader issue of scope and adequacy of the environmental statement' (Appeal Decision Letter, 17 March 1997, para.9). We will look at the uncertainties and the selection process in turn.

## Uncertainties and the RCF

The RCF would, Nirex argued, allow for a complete assessment of the hydrogeology of the site. However, before permission could be granted for the laboratory, Nirex would need to demonstrate that the site had sufficient potential to warrant the expenditure on the laboratory. It was widely considered that if the RCF went ahead then the outlay of expenditure would make it likely that the repository would go ahead. In addition, it was known that Nirex intended to submit a planning application for the repository in 1999, long before the full results from the RCF would be available. Critics argued that the application should not be made until 2005 in order that the RCF could provide the data on which a full evaluation of safety could be made. It appeared that Nirex, in its anxiety to develop a repository, was prepared to neglect the requirement for a rigorous scientific investigation that would enable adequate time for results to be evaluated before any decision to proceed with a full repository proposal was made.

Following the public inquiry into the RCF an internal memo from the Head of Science at Nirex fell into the hands of Cumbria County Council. The memo referred to some detailed aspects of the scientific case for the Sellafield site but also indicated some of the continuing uncertainty surrounding the analyses of the hydrogeological character of the site. The **permeability** of the geological structures at the site was proving to be hard to determine. The memo stated that lower permeability would need to be demonstrated, otherwise, 'I have the feeling that we may struggle to make a case for the site' (10 December 1996). The gas pathways through the rocks also needed to be analysed more clearly so that a report could be generated 'which calculates what we actually think would happen rather than what we think would not happen but which would be a problem if it did'. Furthermore, the then Head of Science 'was concerned that after £200 million the modellers are saying that we are short of datapoints by a factor of 10x or 100x'. The lack of data points and the need for more analysis demonstrated that large areas of inaccuracy and ignorance about the site persisted. More data and analysis would hopefully have overcome these types of uncertainty. More seriously, the uncertainties over safety of the site may well have been related to the unsuitable nature of the geology (a direct result of an inappropriate selection procedure). To understand why, it is necessary to look more closely at the hydrogeology of the site.

> **Permeability** refers to the extent to which substances, like rocks, are able to convey water and gases through them.

The scientific case for the site revolved around interpretations of the geology and hydrogeology of the chosen site: see Figures 1.14 and 1.16. The site is an example of basement under sedimentary cover (BUSC) (shown in Figure 1.10e). The BUSC concept refers to an arrangement of rock layers where sedimentary rocks overlie basement rocks. The flow of water in the basement rocks is slow and

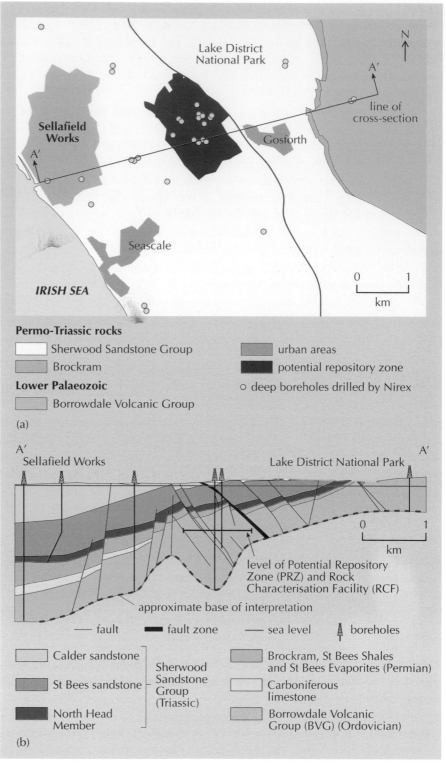

**Figure 1.14** (a) Simplified bedrock solid geology map of the Sellafield area, showing the location of the deep boreholes and the west-south-west/east-north-east cross-section. (b) Cross-section of the Sellafield area along the line shown in (a). Source: Nirex (supplied to the authors), 2003.

water movement into overlying rocks is severely impeded. This would limit the main pathway that radioactive substances might take in migrating from a repository to the surrounding environment. The location of the RCF was to be about 735 metres below the surface in the Borrowdale Volcanic Group of rocks which underlie later sedimentary sandstones and shales dipping westwards towards the coast (shown in Figure 1.14b).

This location was, as Nirex's opponents claimed, actually a variant of the generic BUSC concept portrayed in Nirex publications. An **outcrop** of the basement rock (that is, the underlying rock formation exposed at the surface) meant that there was a potential driving force for the return of water to the surface at a rate which, though slow, might enable radioactive substances to migrate from a repository within a relatively short time-period (see Figure 1.15).

outcrop

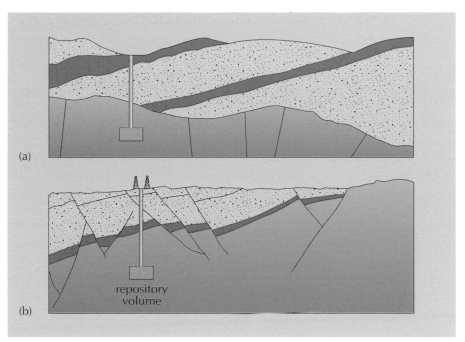

**Figure 1.15** Comparison of the conceptual basement under sedimentary cover with the geology at the site of the RCF: (a) original BUSC concept (as in Figure 1.10e); (b) actual Sellafield basement variant.
*Sources:* Cumbria County Council, 1992, Figure 2.2(a) taken from Nirex Report 71, 1989, and Figure 2.2(b) from Nirex Report 263, 1992.

Moreover, there appeared to be two other features which might increase this risk of hazardous materials reaching the wider environment. One was the apparent presence of very saline water or brine (several times saltier than sea water) to the west of the proposed RCF. This water is of a higher density than fresh water and could prevent the less saline water from flowing down through the basement rocks (see Figure 1.16). The other was the evidence of considerable faulting within the area. Nirex's opponents claimed that faults could provide potential pathways for the rapid transfer of water from the repository to the sandstone above: see Figure 1.14(b).

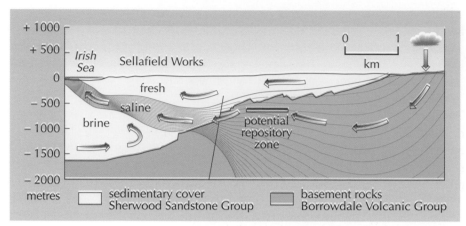

**Figure 1.16** Simplified cross-section through potential repository in Sellafield area (along the line indicated in Figure 1.14a), showing the groundwater flow system and the distribution of the various types of groundwater.
*Source:* Michie, 1998, Figure 14, p.254.

These doubts over the applicability of the theoretical merits of the BUSC concept at this site suggested that not only was the project affected by uncertainties in the form of ignorance (resulting from lack of data), it was also, opponents were arguing, impossible to determine the safety of the site (in other words, there were indeterminacies). Nirex considered that the evidence gained from boreholes and other evidence was sufficient to support its claim that the site held 'good promise as a suitable location for a repository' (Nirex, 1993, p.40). The company recognized that the data, notably on the hydrogeology, were insufficient to justify an application for a repository, hence the decision to seek more exact data through the construction of the RCF. But opponents claimed that the uncertainties were permanent. For them, adding more knowledge through an installation like the RCF would simply expose more uncertainties and would make a satisfactory risk assessment even more unlikely. You should note that we have only focused on the hydrogeological issues. The uncertainties are compounded as soon as we add in human actions (terrorism, future drilling activities), other environmental changes including sea level rises, glaciations and tectonic movements (earthquakes): all of these are either impossible or extremely difficult to predict and illustrate the social and physical complexity of environmental responses.

### Activity 1.8

From what you have read so far, do increases in scientific understanding reduce or add to uncertainties?

### Comment

Both may be the case. Obviously the more we understand, the more certain we become about some things. For example, by gathering more data we may reduce ignorance. But increasing our understanding may also reveal issues of which we were unaware. For instance, as knowledge of the hydrogeology of Sellafield increased, so doubts were cast on its suitability as a potential site for a repository. In other words, with more understanding the indeterminate nature of the risks can become more apparent.

The failure to address many of these uncertainties led the Secretary of State to describe the proposal as 'seriously premature' and 'too optimistic' about its scientific knowledge (Decision Letter, 6, p.xx). The site was also described as more geologically complex than would be expected if it had been chosen to satisfy predominantly scientific criteria. As such, the selection process was under fire.

## Criticizing the selection process

> The UK is ... unique in aiming to have a repository for long-lived ILW waste deep in fractured saturated rock operational by 2010.
>
> (Royal Society, 1994, p.65)

Nirex had chosen a site late in the selection process that would 'not have survived the full series of geological checks in the reduction process if it had been included from the start' (Inspector's Report, Ch.6B.101). It had clearly been chosen because of its proximity to the nuclear industry's base and the promise of local support which might well prove 'at the most a transitory advantage in relation to such a long-term project' (ibid., 6B.104). The company conceded that 'there are likely to be radiologically better sites available around the UK' (Decision Letter, 7). In choosing Sellafield, Nirex was forced to prove a scientific safety case in a geological setting that its critics considered was complex in terms of its hydrogeology, faulting and seismicity.

We have already noted that the repository was scheduled to open before research from the RCF could be fully retrieved and assimilated to support the case for a repository. Some experts argued that on both scientific and economic grounds it might be better and cheaper to proceed more slowly. Scientifically, time would allow the research process to reduce some uncertainties and rule out other options where uncertainties became more apparent. Economically, there are two arguments. First, it would make sense to delay expenditure until a more promising case could be made and so avoid costly mistakes. The second reason is that it might make economic sense since greater value for money may be achieved by setting aside financial provision but deferring the actual investment. Expressed in today's prices the future costs could be considerably lower, a process known as '**discounting**' (see Box 1.5).                    discounting

Years later, Nirex conceded that 'the project was driven too quickly and was focused on getting to the end of the project as soon as possible' (Nirex, 2001, p.9). The site selection process had not been transparent and scientific information was not readily available in the public domain, thereby adding to the perception of a lack of openness on the part of the company.

## Box 1.5  Discounting the future

This is a means of making sufficient financial provision today for projects which, like radioactive waste repositories, must be developed at some future date. It is assumed, with some confidence, that economic growth will continue and that, in consequence, a project, in terms of today's prices, will be cheaper in the future. It is necessary to calculate how much investment is needed now to ensure sufficient finance will be available in future.

The amount invested now will depend on two variables. One is the discount rate, which is the rate of depreciation of money values calculated as an annual percentage (a rate of 10 per cent would mean that the monetary value of something would reduce by one tenth in a year's time). The other variable is the time over which the calculation is made. At very long timescales projects become very cheap by today's prices. In simple terms if the cost of building a repository at today's prices is £1 billion, then, in 100 years' time it might be expected that this cost would, in real terms, be significantly lower. In fact, the cost, using a 5 per cent discount rate, would be only £87.2 million. By delaying the building of a repository the present generation is discounting the future. This raises ethical questions about how we value the future and we shall return to this in the final part of this chapter.

For more on discounting, see Pearce et al. (1989) and Chapter 6 in this volume.

In practice the selection of Sellafield was another example of the DAD strategy. Following a change in government in 1997 a number of reports focused upon the problem of dealing with radioactive wastes. A general theme emerged on the need for openness and transparency in order to generate public trust and support. Hitherto, solutions had been decided by experts and defended publicly but these had ended in defeat for Nirex. In future it was intended that policy options would emerge from public debate informed by experts. This approach envisaged the possibility of a consensus that could transcend the short-term rhythms of political conflict in order to develop a long-term strategy to deal with a problem extending far into the future. In the final subsection we will investigate what this search for consensus involves.

## 3.4  The search for consensus

As you saw in Section 2 a number of commentators – from scientists like May to sociologists like Irwin – are urging a shift towards deliberative forms of decision-making and policy design. Deliberative approaches were described as means to generate active debate on policy issues with the aim of generating a working consensus. These methods are widely promoted as improvements to top-down forms of policy-making. Here we will explore in a little more detail what these approaches involve before assessing their ability to answer our third question of whether or not they can take adequate account of future generations' needs and whether, therefore, they are consistent with the principle of sustainable development. The basic features of a consensus-orientated approach are set out in Box 1.6.

## Box 1.6   Components of consensus

| Component | Advantage |
|---|---|
| Adequate time to take necessary decisions | avoids premature decision-making |
| Wider peer review | encourages credibility of science |
| Early involvement of those affected by decisions | helps to ensure accountable and legitimate decisions |
| Deliberative and accessible decision-making | encourages wide participation and social consent |
| Greater openness and transparency in decision-making | engenders greater trust in science and institutions |
| Commitment to equity | applies concepts of compensation to communities affected by decisions |

The search for consensus may well prove important in developing successful responses to environmental issues: these approaches can be more responsive (in part by taking into account more uncertainties) and can produce results which have greater legitimacy and chance of success. It is important to recognize that consensus implies consent for a strategy; it does not signify total agreement.

### Activity 1.9

With the help of Box 1.6, in what ways did Nirex's handling of the RCF depart from a consensus-based approach?

### Comment

The main departures were: a tendency to rush the process; a failure to make scientific findings public; a strategy of Decide–Announce–Defend; a failure to involve the wider public; a secretive approach to decision-making and a failure to disclose uncertainties; and site selection based primarily on political criteria which did not include a consideration of equity issues. These all contributed to the selection and ultimate refusal of a site that was considered to be inappropriate by the Planning Inspector and Nirex's critics.

We shall explore the consensus-based approach in more detail, taking each of the components in Box 1.6 in turn.

First, the time needed to reach consensus is often long. In the case of the RCF there was insufficient time for a measured scientific programme. This experience suggests that adequate time must be given for appropriate scientific investigation. However, this does not mean that inordinate delay is acceptable. In the fallout case discussed earlier, it was evident that time was of the essence and rapid decisions were needed.

Second, the credibility of any scientific research depends on peer review, that is the examination of the work by suitably qualified people. In the Nirex case, while much of the scientific work was acknowledged to be of high quality, there was pressure for results which led, in some cases, to an optimistic interpretation to be put on results and to a tendency to marginalize the uncertainties. It is important to make sure that the review process does not lead to a closed shop of mutual approval, and that outsiders are encouraged into the expert deliberation.

Third, early involvement of those most likely to be affected by decisions is necessary in order to avoid the mistrust and resentment that Nirex encountered in its early attempts to find sites. The experience suggests that, 'unless people believe that the right policy has been adopted for the right reasons, they will strongly resist its implementation' (Nirex, 2001, p.41).

Fourth, it is important to engage a wider public since, as the Cumbrian sheep-farmers example has shown, lay understandings are important and, as the RCF case suggests, cannot be taken at face value or for granted. Nirex accepted the lesson that to achieve public acceptability it is necessary 'to stimulate informed discussion, and to involve as many people and groups as possible' (Nirex, 2001, p.7). However, it must also be recognized that involving people in **deliberative decision-making** (see Box 1.7) can be difficult, given that many people find the time needed to act on panels or deliberative groups highly restricting. This may be particularly the case for those on limited incomes.

deliberative decision-making

Fifth, the tendency to keep things secret and to be parsimonious with the release of information was characteristic of the experts in both the fallout and the RCF cases. This lack of openness was later recognized to be counterproductive. A review of radioactive waste policy by a committee of the House of Lords argued that there should be 'a general presumption of disclosure, openness and transparency. This is a necessary condition for achieving public trust' (House of Lords, 1999, p.41).

Finally, there may be resentment against decisions that are perceived to be unfair. This was so in both the studies. In the fallout case, farmers felt unfairly treated by decisions. At Sellafield, people were fatalistic and to an extent victimized. In order to reduce this tendency, forms of compensation will be necessary in recognition of the risks that are borne by a community on behalf of the wider society.

## Box 1.7   Consultation techniques

The conventional (top-down) method of issuing a consultation paper and inviting responses is being supplemented by methods that supply information, stimulate interest and seek to get feedback and gauge opinion. These include interactive websites, newspaper adverts, free phone lines, workshops, telephone surveys and focus groups. There are also more sophisticated deliberative techniques which engage small groups (no more than 20) in dialogue to provide an informed opinion on a subject. These may be quite small groups focusing on specific local issues. They include citizens' juries which examine witnesses and reach a consensus; community advisory groups which consider an issue over several meetings; and interactive panels which are representative standing panels providing views on issues such as health. Consensus conferences consist of a panel of lay people who take a particular subject of national concern (such as radioactive waste) and call on and question expert witnesses in public to achieve a set of policy recommendations. Much larger groups are recruited to form research panels (500–5,000) and deliberative opinion polls to see how public views develop as they gain a deeper understanding of the issue. Lastly, stakeholder dialogues (including some on nuclear issues run by British Nuclear Fuels Limited) have been initiated to see how far a consensus can be achieved by participants with very different interests and frequently opposed attitudes.

Commitment to equity and compensation become difficult as soon as the relevant community is extended beyond the present generation. We will conclude on this issue of equity by considering the interests of future generations. Focusing on their needs allows us to judge the sustainability or otherwise of current approaches to environmental responses.

## Taking future generations into account

Equity has both spatial and temporal dimensions. Spatial equity applies to the uneven distribution (or inequality) of hazard between places. We have considered ways in which this might be dealt with elsewhere in the series **(Blowers and Elliott, 2003; Maples, 2003)**. In terms of time, equity draws attention to the uneven distribution of hazards between generations. It is difficult to deal with such issues in the consensus approaches that we have explored here mainly because it is impossible to invite future generations to the public meetings. One way around this is to use a number of principles to guide action. Thus, the principle of **intergenerational equity** assumes that potential impacts on health    intergenerational equity to future generations should be no greater than those which are acceptable today. This is related to the idea of sustainable development. In terms of radioactive waste, this can be interpreted in two ways. One is a tough but defensible principle that 'future generations should not have to bear the costs and consequences of actions which were of benefit mainly to present and previous generations' (House of Lords, 1999, p.89). Put simply, it is inequitable for us to benefit from nuclear

electricity while consigning the problem of nuclear waste to the future. The other interpretation is the principle of freedom of choice – that we should not compromise the ability of future generations to meet their own needs. In other words, we should, as far as possible, avoid foreclosing options for the future.

### Activity 1.10

How might these principles help us decide whether to proceed with a repository now or to delay and undertake further research?

### Comment

By dealing with the problem of radioactive waste now, we may avoid imposing economic costs on future generations but may transfer some risks. On the other hand this approach may forfeit the possibility that a strategy of wait and see might offer the freedom of choice to develop better solutions later that will minimize future risk. Of course, by continuing to store wastes above ground we leave options open but also impose costs on the future.

There is a further problem in how we conceive of the future. We tend to think in terms of what may be called 'cultural time', that is time which we can imagine on the basis of our experience. Research has indicated that our forward perception of time is relative to self, family or community and extends to little more than two generations and at most a hundred years (Duncan, 2002). But, cultural time is but the blink of an eye compared to the period, of at least 100,000 years, over which management of solid radioactive wastes must prevent the migration of radioactive substances into the accessible environment. Duncan's research suggests that our perception of time leads us to believe that geological disposal will fail within a relatively short time and that claims that isolation of waste can be achieved for up to 100,000 years 'are incomprehensible and therefore will be rejected' (ibid., p.77). In any case, such geological timescales are well outside any credible limits of institutional survivability. Indeed, it may be argued that climatic and geological uncertainties render statistical calculations of risk beyond even 10,000 years

The contrasting scales of time are discussed in **Reddish (2003)**. **Bingham and Blackmore (2003)** discuss cultural perceptions of risk and credibility of institutions in charge of risky activities.

unreliable (Table 1.1). Thus, the regulatory standard of less than a $10^{-6}$ (i.e. one in a million) per year risk of an individual suffering a serious health effect from a release of radioactivity from a repository is not only meaningless to the average member of the public but, beyond a certain time, becomes scientifically and socially incredible.

Table 1.1   Time-frames of past and future possible events.

| Years | Historical | Future | Approximate half-lives of some radionuclides* for comparison |
|---|---|---|---|
| 100 | Discovery of radioactivity | Enhanced greenhouse effect | iodine-131 (8 days) caesium-137 (30 yrs) |
| 1 000 | Norman conquest | Large ecological changes (e.g. estuarine development) | |
| | Egyptian pyramids | | |
| | | Exhaustion of mineral and fossil fuel resources? | carbon-14 (5,730 yrs) |
| 10 000 | | | |
| | Last glaciation of northern Europe | | plutonium-239 (24,000 yrs) |
| | Use of fire and tools a by humans | | |
| | | Next glaciation | |
| 100 000 | Emergence of Neanderthal man | Time between major glaciations | |
| | | | technetium-99 (213,000 yrs) chlorine-36 (301,000) |
| 1 000 000 | Emergence of *Homo sapiens* | Stable geological formations remain largely unchanged | |
| | Postulated evolutionary branching between humans and apes | | |
| 10 000 000 | | Appearance of new families of species? | iodine-129 (15.7 million yrs) |
| 100 000 000 | Dinosaurs populate the Earth | Large-scale movements of continents (thousands of kilometres) | |
| 1 000 000 000 | Appearance of multicellular organisms | Significant probability of 'nearby' supernova having occurred | |
| | | Increase in solar intensity sufficient to destroy life on Earth | uranium-238 (4,500 million yrs) |
| | Age of the Earth | Sun becomes red giant | |

*Notes:* *A radionuclide is any radioactive atom of an element identified by the number of neutrons and protons in its nucleus, and its energy state.

*Source:* Nirex, 1995, p.4.

ROLLING PRESENT

The scientific and social uncertainties that increase as we extend the time horizon, as well as the cultural propensity to focus on the present and the immediate future, suggest that our capacity to take moral responsibility for the future is limited. Pragmatically speaking we must let the far future take care of itself. The somewhat rigid choice between act now or wait-and-see is not especially helpful in guiding our actions. In their place a less rigid 'principle of retrievability' can be used to make sure that options are not foreclosed. Looked at this way, decision-making can be seen as a continuing process in which present and future are linked. Each generation has the ability to change course, taking into account changing environments, technologies, scientific understandings and social values.

## Summary

In this section we have looked in detail at the failure of conventional approaches to the management of radioactive wastes. The failure has been blamed upon a policy process that has been unable to secure consensus or legitimacy. Alternative ways forward have been discussed and their limitations, notably with respect to future generations, considered.

## 4    Conclusion

In this chapter we have discussed how environments and people respond to changes that carry a risk of harm. You should now appreciate that environments respond in complex ways that involve all manner of physical processes, from atomic to planetary in scale. Responses of environments involve living organisms and people. Taken together, these physical, biological and social responses are often difficult to predict. From the pathways of caesium-137 in soils, to the movement of water in sedimentary rocks, and from the marketing of sheep to the actions of terrorists or future governments, uncertainties come to characterize our ability to assess risks of present and future hazard. Such complexities and the uncertainties that accompany them make social responses to potentially hazardous changes problematic. Not knowing when or how to act has been a feature of both cases that have been discussed in this chapter. Nevertheless, the imperative to act has led to a discussion of the forms of decision-making and policy that are either conventionally used or are offered as alternatives. In both cases policy-makers have been criticized for expert-led, top-down approaches to decision-making and more deliberative, consensus approaches have been discussed as alternatives. At their best these alternatives offer more responsive forms of governance which can take greater account of uncertainties. We have noted some limitations to these alternatives, not least the difficulties in incorporating the needs of future generations. In the end we have suggested that pragmatic responses, which will by no means be perfect but which can be responsive to the dynamic environments in which we live, may offer the best ways forward. As we noted at the outset, these dynamic environments include changes in technological, economic and political circumstances. In the following chapters you will read more about each of these components of environmental responses.

# References

Bingham, N. (2003) 'Food fights: on power, contest and GM' in Bingham, N. et al. (eds).

Bingham, N., Blowers, A.T. and Belshaw, C.D. (eds) (2003) *Contested Environments*, Chichester, John Wiley & Son/The Open University (Book 3 in this series).

Bingham, N. and Blackmore, R. (2003) 'What to do? How risk and uncertainty affect environmental responses' in Hinchliffe, S.J. et al. (eds).

Blackmore, R. and Barratt, R.S. (2003) 'Dynamic atmosphere: changing climate and air quality' in Morris, R.M. et al. (eds).

Blowers, A.T. and Elliott, D.A. (2003) 'Power in the land: conflicts over energy and the environment' in Bingham, N. et al. (eds).

Blowers, A.T. and Smith, S.G. (2003) 'Introducing environmental issues: the environment of an estuary' in Hinchliffe, S.J. et al. (eds).

Burgess, J. (2003) 'Environmental values in environmental decision making' in Bingham, N. et al. (eds).

Cumbria County Council (1992) *A Review of the Geology and Hydrogeology of Sellafield*, Interim Technical Appraisal Report ITA/4, June, Cumbria County Council.

Drake, M. and Freeland, J.R. (2003) 'Population change and environmental change' in Morris, R.M. et al. (eds).

Duncan, I. (2002) 'Disposal of radioactive waste: a puzzle in four dimensions', *Nuclear Energy*, vol.41, no.1, February, pp.75–80.

Gale, H.J., Humphreys, D.L.O. and Fisher, F.M.R. (1964) 'Weathering of caesium-137 in soil', *Nature*, vol.201, no.4916, pp.257–61.

Gould, P. (1990) *Fire in the Rain: The Democratic Consequences of Chernobyl*, Cambridge, Polity Press.

Hinchliffe, S.J., Blowers, A.T. and Freeland, J.R. (eds) (2003) *Understanding Environmental Issues*, Chichester, John Wiley & Son/The Open University (Book 1 in this series).

House of Lords (1999) *Management of Nuclear Waste*, Select Committee on Science and Technology, Session 1998–99, Third Report, London, HMSO.

Irwin, A. (1995) *Citizen Science: A Study of People, Expertise and Sustainable Development*, Routledge, London.

Irwin, A. (2001) *Sociology and the Environment*, Cambridge, Polity Press.

Maples, W.E. (2003) 'Environmental justice and the environmental justice movement' in Bingham, N. et al. (eds).

May, R. (1997) *The Use of Scientific Advice in Policy Making*, Office of Science and Technology, DTI, London, available at http://www.dti.gov.uk/ost/ostbusiness/policy.htm

Michie, U.M. (1998) 'Deep geological disposal of radioactive waste: a historical review of the UK experience', *Interdisciplinary Science Reviews*, vol.23, no.3, pp.242–57.

Morris, R.M. (2003) 'Changing land' in Morris, R.M. et al. (eds).

**Morris, R.M., Freeland, J.R., Hinchliffe, S.J. and Smith, S.G. (eds) (2003)** *Changing Environments,* **Chichester, John Wiley & Son/The Open University (Book 2 in this series).**

National Enviornment Research Council (1992) *Twenty Fifth Scientific Report of the Institute of Terrestrial Ecology 1997–1998,* (eds T.M. Roberts, K. Threlfall and J. Sheail), Centre for Ecology and Hydrology, Swindon, NERC.

Nirex (1987) *The Way Forward,* Harwell, United Kingdom Nirex Limited.

Nirex (1989) *Deep Repository Project: Preliminary Environmental and Radiological Assessment and Preliminary Safety Report,* Harwell, United Kingdom Nirex Limited.

Nirex (1989) *Nirex Report 71,* Harwell, United Kingdom Nirex Limited.

Nirex (1992) *Nirex Report 263,* Harwell, United Kingdom Nirex Limited.

Nirex (1993) *Scientific Update 1993: Nirex Deep Waste Repository Project,* Harwell, United Kingdom Nirex Limited.

Nirex (1994) *Rock Characterisation Facility: Longlands Farm, Gosforth, Cumbria – A Summary of the Environmental Effects,* Harwell, United Kingdom Nirex Limited.

Nirex (1995) *Science Report: Post-closure Performance Assessment – Treatment of the Biosphere,* Report no: S/95/002, May, Harwell, United Kingdom Nirex Limited.

Nirex (2001) *Report on the Nirex Internal Inquiry January–December 2000,* Harwell, United Kingdom Nirex Limited.

Nirex (no date) *Going Forward: The Development of a National Disposal Centre for Low and Intermediate-level Radioactive Waste,* Harwell, United Kingdom Nirex Limited.

Pearce, D., Barbier, E.B. and Markandya, A.(1989) *Blueprint for a Green Economy,* London, Earthscan.

Phillips, L. (1995) *Multi-attribute Decision Analysis for Recommending Sites to be Investigated for their Suitability as a Repository for Radioactive Wastes,* Proof of Evidence on behalf of UK Nirex Ltd., IN431, October.

**Reddish, A. (2003) 'Dynamic Earth: human impacts' in Morris, R.M. et al. (eds).**

Royal Society (1994) *Disposal of Radioactive Wastes in Deep Repositories,* London, Royal Society.

Wynne, B. (1990) 'Sheepfarming after Chernobyl; a case study in communicating scientific information' in Bradby, H. (ed*.) Dirty Words: Writings on the History and Culture of Pollution,* London, Earthscan.

Wynne, B (1992) 'Uncertainty and environmental learning: reconceiving science and policy in the preventive paradigm', *Global Environmental Change,* vol.2, no.2, pp.111–27.

Wynne, B. (1996a) 'Misunderstood misunderstanding: social identities and public uptake of science' in Irwin, A. and Wynne, B. (eds) *Misunderstanding Science? The Public Reconstruction of Science and Technology,* Cambridge, Cambridge University Press, pp.19–46.

Wynne, B. (1996b) 'May the sheep safely graze? A reflexive view of the expert-lay knowledge divide' in Lash, S., Szersynski, B. and Wynne, B. (eds) *Risk, Environment and Modernity: Towards a New Ecology,* London, Sage, pp.44–83.

Wynne, B., Waterton, C. and Grove-White, R. (1993) *Public Perceptions and the Nuclear Industry in West Cumbria,* Centre for the Study of Environmental Change, Lancaster, Lancaster University.

# Chapter Two

# Design for urban environments

Rod Barratt

## Contents

# 1    Introduction

(The view) is quite spectacular and always changing. ... At night we see cities all lit up in populated parts of the world. It's quite amazing to see how many people live down there and the effect they're having on the environment and the land we live on ...

It's a cause for concern. Because, since my first flight in 1990 and this flight I've seen changes in what comes out of some of the rivers, I've seen changes in land usage and we see areas of the world that are being burned for clearing the land, so we're losing lots and lots of trees ...

So yes, it's a cause for concern. We have to be very careful how we treat this good Earth we live on.

This extract from an interview with the commander of the International Space Station was broadcast late in 2001 on BBC Radio 4. His concerns are illustrated by the two satellite images of the Chinese city of Shenzhen (Figure 2.1). China has one of the fastest growing economies in the world, and cities like Shenzhen have grown dramatically in recent years.

Notice how, in the images of Shenzhen, urban development has transformed the landscape. New structures have appeared off the southern coast, and roads grow less distinct as development occurs along their length.

In this chapter we will focus on our human tendency to accumulate in groups – in urban areas. We will explore, to a degree, how urbanization can accommodate the needs of growing human populations efficiently. This book is about responding to environmental issues and, in this chapter, we focus on technological responses. We will explore some ways in which we can be proactive, through using resources efficiently and designing with environments in mind.

(a)                                                    (b)

**Figure 2.1** The city of Shenzhen in China has grown dramatically since 1988: (a) Shenzhen in 1988 and (b) Shenzhen in 1996. These satellite images were taken from the LandSat website.

In addressing questions that focus on managing environmental problems and making desirable and sustainable environmental futures, we aim to:
- look at ways of addressing the risks and uncertainties of environmental futures at the level of the household and 'urban community';
- suggest ways of making informed decisions on urban development;
- emphasize the responsibility for 'environmental management' at different levels including at the individual level;
- illustrate the hierarchy for dealing with 'wastes';
- outline some approaches for improving 'eco-efficiency'.

We will also help you develop some skills with wider applications, including:
- representing complex problems diagrammatically;
- interpreting graphical and numerical data;
- integrating scientific and social approaches in environmental management.

# 2   The scale of concern

A key to effective action in responding to urban environmental problems involves an understanding of the scale of the issues. This understanding will help us to collect information at the appropriate scale, identify resources and stakeholders relevant at that scale, and take appropriate action at the appropriate level. But before we look at different scales in cities, we need to understand something of the scale of contemporary cities.

## Urbanization

**Activity 2.1**

Estimates suggest that the total population in developing countries grew from 1.7 billion in 1950 to approximately 4 billion in 1990, while their urban population increased during the same timespan from 286 million to 1.514 billion. Compare these changes by calculating the respective growth ratios of developing countries first and then their urban populations. For the answer, go to the end of this chapter.

Since 1990, the proportion of the world's population living in urban areas has continued to grow faster than other demographic statistics. It reached 47 per cent in 1998 and predictions suggest 55 per cent by 2015. In Europe and the Americas 80 per cent of people are already city dwellers, but the biggest urban growth rates up to the year 2025 will be in Asia and Africa. If predictions come true, then almost two-thirds of the world's population will be living in urban areas by that time.

Population growth and cities are discussed in **Drake and Freeland (2003)**.

Urban population growth occurs through both migration and natural growth. More and more of us live in so-called 'megacities' (cities with over ten million inhabitants, see Figure 2.2) In 1950, the five biggest cities were New York, Tokyo, London, Paris and Moscow. Today the European ones have been replaced by Sao Paulo, Mumbai (Bombay) and Mexico City, of which more later. By 2015, the top five may be Mumbai, Lagos, Shanghai, Jakarta and Tokyo.

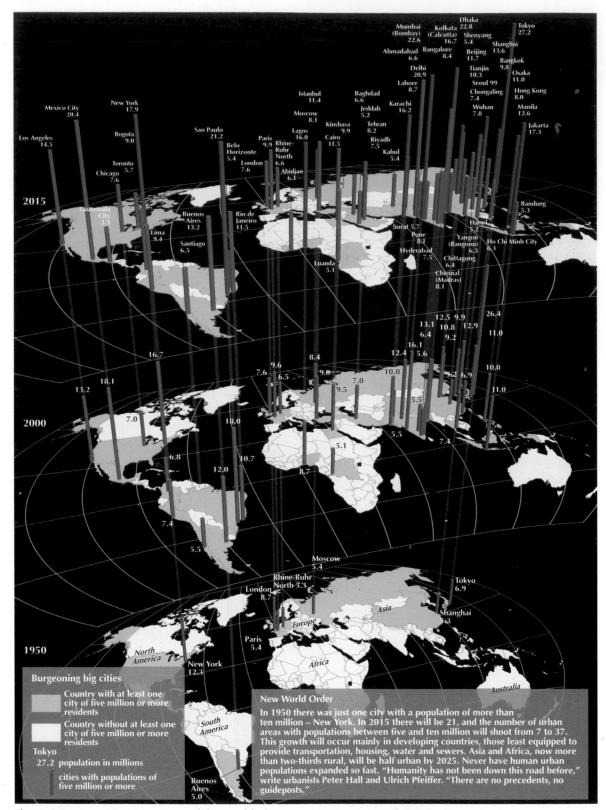

**Figure 2.2** Megacities. Locations of the world's biggest cities.

Cities grow around activities that are best carried out centrally, such as government and manufacturing. Industrialization imposes pressures on the environment and human health from several sources, including the extraction and consumption of raw materials for manufacturing, increased energy demand and emissions of pollutants (**Reddish, 2003**). Strong legislation in most developed countries has controlled the worst pollution, although problems still remain, notably climate change. In less developed countries, pressures for economic growth are intense, and regulations and enforcement may be more lax. Resource consumption remains prominent across the world, especially in the wealthiest countries, which consume vast quantities of raw resources to support their quality of life (**Reddish, 2003**). Cleaner and more efficient industry could greatly influence health and the environment in and around cities and address our key questions on approaches for managing environmental problems as well as providing opportunities for sustainability. Reducing environmental impact at the household level could also be worthwhile. See from Figure 2.3 how, for the UK, the 30 per cent growth in household numbers was matched by the growth in several key indicators of environmental impacts. Water and energy consumption have followed the increasing use of appliances in the home, despite improvements in their energy efficiency. Domestic waste grew by a third between 1983 and 1999 – twice as much as household numbers. We will look at some approaches for managing environmental issues at the household level shortly.

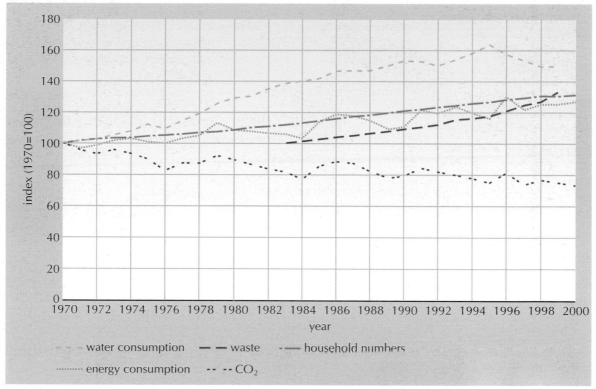

**Figure 2.3** The environmental impact of households in the UK, 1970–2000.
*Source*: Department for Environment, Food and Rural Affairs, *The Environment in your Pocket, 2001*.

Cities are encouraged by: the development of services such as banking and the accumulation of labour skills; opportunities for social, recreational and cultural activities; and the value of cities as centres for communications. The globalization of industry and trade stimulates the growth of megacities. So does the mechanization of farming, which reduces demand for labour in the countryside and releases large numbers of people in the most heavily populated countries to migrate to cities in search of work. This places even greater pressures on the trade in agricultural products.

## Suburbanization

The second half of the twentieth century revealed a growing trend – suburbanization. This broadens the scale of urban impact. While suburbs have a long history, cities are increasingly diffused in both developed and developing countries. Jakarta is growing less fast than the wider urban zone stretching the length of northern Java. A network of secondary cities is growing on the Brazilian coast between Rio de Janeiro and São Paulo. Calcutta has stopped growing while new cities emerge in the hinterland of West Bengal. **Urban sprawl** is a growing environmental problem in terms of traffic, resource consumption, resulting pollution and so on.

urban sprawl

The rate and scale of change – over 60 million people join urban populations each year – can strain the capacity of governments to provide even the most basic services to urban residents. Many in developing countries have little or no access to adequate water, sanitation and waste management, and more than a billion urban inhabitants breathe air polluted to harmful levels (see Box 2.1).

### Box 2.1  Mexico City

Mexico City is situated in a basin surrounded by high mountains, relies heavily on cars and buses, many of which are poorly maintained, and has a large number of polluting industries, combined with a predominantly dry and sunny climate. These are factors necessary for photochemical smog formation (**Blackmore and Barratt, 2003**) and combine to make Mexico one of the world's smoggiest and most unhealthy cities. Later we see some approaches that are improving the environment of Mexico City.

In 2001 only 23 per cent of Mexico's water supplies were adequately treated, with the area surrounding the capital having the most contaminated aquifers. Some 30 to 50 per cent of water is lost through leaks.

Although urbanization can put environmental quality and human health at risk, cities are not all bad. They can bring benefits, but can the benefits of cities justify the environmental impact of those cities, particularly in terms of the energy and resources they consume? They have potential advantages: they drive economies, and economic growth can combat poverty, which is a significant threat to environmental quality. Most of the fastest growing urban areas in the developing world are in countries with the fastest growing economies like China. High population densities can minimize resource use and energy consumption – potentially by developing mass-transit systems to supplement car use, for example, or living in resource-effective residences close to where we work or study. You will come across other potential advantages of urban living as you read on. In the following we will tend to focus on the hazards and risks that cities contribute towards and their possible amelioration through better design at micro and macroscales.

## 2.1 The complexity of the issue

With the risk of over-generalization, we could represent urban issues as existing in three dimensions: the natural, the built and the socio-economic (see Figure 2.4).

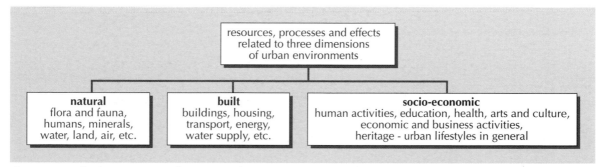

**Figure 2.4** Urban environmental issues can be seen as existing in three dimensions: the natural, the built and the socio-economic.

Urban environments involve complex interactions between these dimensions and we must understand their interdependency if we want to develop coherent and sustainable policies to change future urban environments.

Another model of urban environmental issues involves a consideration of *resources*, the *processes* that convert these resources into various products and services, and the *effects* of these processes, whether negative or positive (see Table 2.1).

**Table 2.1**   A list of the resources, processes and effects involved in urban environmental issues

| resources | | human; sunlight; land; water; timber; cereals; minerals; electricity; fuels; finance; intermediates; recyclables. |
|---|---|---|
| processes | | manufacturing; transportation; construction; migration; population growth; habitation; community services (education, health, etc.) |
| effects | **positive** | products; value-addition; increased knowledge base/education; access to better services. |
| | **negative** | pollution (air, water, noise); waste (municipal, sewage, industrial); congestion and overcrowding; loss of agricultural land, etc. |

This focus on processes and their effects is similar to the source–pathway–effect model used as a risk assessment tool in decision making by regulators. Looking at urban processes in this way can help us to think more clearly about managing risks and environmental hazards. The concept of risk has two elements: the likelihood of something happening and the consequences if it happens. For a risk to exist there must be an identified or plausible linkage between the source of harm and the receptor (see Figure 2.5).

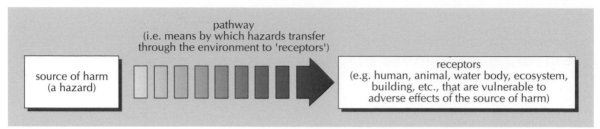

Figure 2.5 The source–pathway–effect model of risk assessment.

In technological discussions, the distinction that is usually made between risk and hazard is that hazard is an intrinsic property with the potential to cause harm. Risk takes account of the hazard and its likelihood of occurrence (**Bingham and Blackmore, 2003**).

You met the idea of pathways in Chapter One. Recognition of the pathways by which hazards are conveyed can help us identify and manage risks. For example, public health risks from water supplies contaminated by sewage are reduced by an urban infrastructure that provides sewerage systems to treat wastes, while treatment of water supplies eliminates harmful micro-organisms.

### Activity 2.2

Try to apply the source–pathway–effect model to some risks around you to explore its implications. For example, when cooking, you may pick up a hot pan using oven gloves. These separate the hazard (the hot pan) from the receptor (your hands).

In addition to understanding causes, or sources, their pathways and their effects, it is useful if we are to manage risks effectively, to understand how people think about and act on risk knowledge. Figure 2.6 shows some hazards ranked according to their potential impact on people and their time of onset. Notice that climate change, primarily caused by our use of energy, gives rise to the highest expected human impacts, but takes the longest time to do so and so may be of limited concern in people's everyday lives.

As this example suggests, the time and scale of different issues can influence our perceptions of the risks associated with them and especially by how much we hold them in 'dread'. Something we know little about can also be perceived as of high risk. These two criteria for risk acceptability (knowledge and dread) are two issues that influence our perceptions, and have been used in a method for grouping

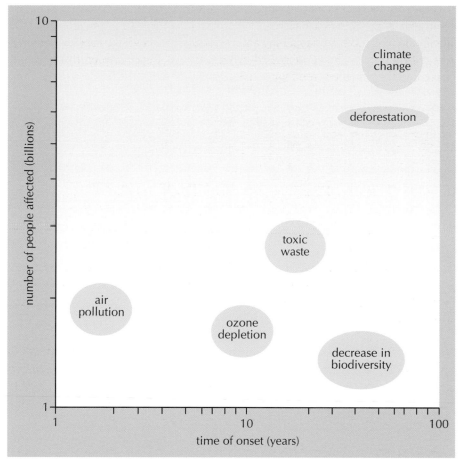

**Figure 2.6** Ranking of environmental risks. This illustrates issues of time and space and the need for intergenerational equity. Note: the axes use a logarithmic scale to accommodate greater scales.

hazards. Under this scheme, a well-established hazard that is observable, known to those exposed to it, has an immediate effect and has a risk that is scientifically measurable, is given a high score on the 'known' scale. Factors that contribute to the 'dread' scale are the seriousness of the hazard, how personal are its effects, the degree of control available to individuals, the scale of catastrophe, equity of hazards between different groups, the risk to future generations, voluntary or involuntary involvement and the potential for reduction. Various technological hazards are grouped according to knowledge and dread in Figure 2.7.

From Figure 2.7 we see that cycling, motor cars and domestic appliances are regarded as having known hazards with low dread, and appear in the bottom left quadrant. In comparison, many sources of power (nuclear power in particular) are less well known and figure more on the dread dimension, so they appear in the upper right quadrant. The public demands greater regulation of those risks that have higher 'dread' and 'unknown' factors. However some risks may be outside our normal scale of perception, especially if they relate to new issues. It is

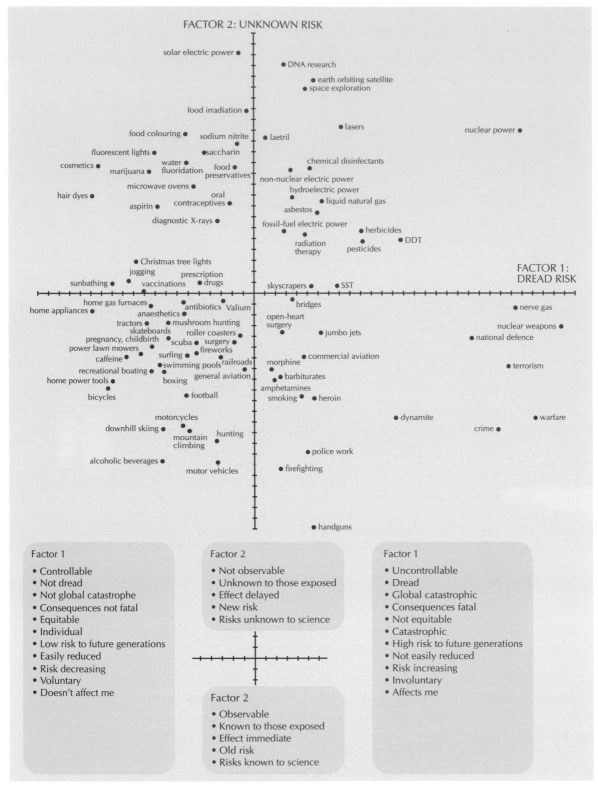

**Figure 2.7** Hazards grouped along dimensions of dread and knowledge.
*Source:* The Proceedings of the Royal Society Series A, 'Mortality in relation to smoking'.

important to understand these elements of risk and hazard and the complexity of urban processes in order to judge the ways in which we organize our cities.

In the next section we look at some ways of addressing urban environmental hazards by changing resource use and processes. Before doing so, we can end this section by looking at some of the tools and approaches that are available to environmental managers.

## 2.2  Some tools we can use

We have a variety of tools for environmental management. In the past, we commonly regulated behaviour through what are known as '**command and control**' approaches – setting standards for water quality, discharge consents and so on. These set a baseline for the minimum performance expected, but could be costly and difficult to enforce. Subsequently, **market-based incentives**, such as subsidies, or taxes, gave financial incentives for industry to adopt less polluting behaviour or technology (Chapter Three looks at the detail of some market-based approaches). **Information**, too, is a powerful tool for facilitating change. Public disclosure of environmental performance encourages pollution reduction while public education campaigns that explain links between environmental hazards and human health can influence individual behaviour.

command and control

market-based incentives

information

We also need other tools. Urban planners need to understand the potential health impacts of projects. Environmental impact assessments (EIAs) are now commonplace, but techniques for assessing risk must be broad and incorporate ecosystem impacts that clarify where long-term or indirect risks might emerge. Biological indicators, such as plant, rodent and insect populations, as well as monitoring of environmental and climatological changes may be useful. Tests may show that a product under development is not directly toxic to humans, but if environmental harm is ignored, over the long term we may be exposed to new health risks from environmental change. (A good example of this is the case of CFC (chlorofluorocarbon) production, see Box 2.2 and **Blackmore and Barratt, 2003.**)

### Box 2.2  Risks from a short-term view of human safety

Invented in 1928 by Thomas Midgley, Jr., CFCs were regarded as miracle compounds because they made many everyday conveniences a reality. They were welcomed as a breakthrough – being non-toxic and with many uses. They were efficient refrigerants, making household refrigeration possible and much safer than using previous refrigerants. They were excellent coolants in air conditioning systems, propellants for aerosol sprays and cleaning agents in the manufacture of computers. It may be no exaggeration to say that much of the improvement in lifestyle from the mid-twentieth century on was made possible by CFCs.

Only later did we find out the harm they could cause. In 1974, after millions of tonnes of CFCs had been manufactured and used, chemists F. Sherwood

Rowland and Mario Molina began to wonder where all these CFCs ended up. The reason CFCs were so effective was because they were chemically stable – nothing in the troposphere caused them to break apart into their constituent atoms of chlorine, fluorine and carbon. Rowland and Molina suggested that ultraviolet light in the stratosphere could break them down releasing chlorine that could react with ozone at stratospheric altitudes (Molina and Rowland, 1974). This thin band of ozone protects all life on Earth from short UV radiation (**Blackmore and Barratt, 2003**) and destruction would expose life on Earth to great risk.

Sparked by these findings, scientists, policy makers, environmentalists and industry embarked on an often contentious debate about the issues of ozone depletion. Many remained unconvinced of the danger until the mid 1980s, when a severe depletion of ozone appeared above Antarctica, dramatically revealed since then through satellite images (Figure 2.8).

Thanks in part to this realization, in 1987 fifty-seven industrial nations signed the Montreal Protocol (the first global environmental agreement) starting a phase-out of CFCs and other ozone-depleting chemicals. Today, several CFC substitutes are in use, although it will be many years before the effects completely disappear.

Despite opposition from the chemical industry, squabbling in the scientific community and the challenge of devising a global agreement to satisfy the requirements of both developed and developing nations, the world ultimately agreed to address a clear environmental risk (see Chapter Three). Achieving international consensus and concerted action on global warming is more problematic, however, because banning CFCs involved only one basic family of industrial chemicals, while global warming is more complex. That issue is addressed in the next chapter and Chapter Five.

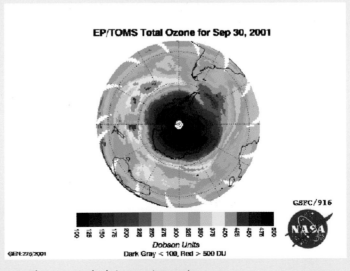

Figure 2.8 The 'ozone hole' over Antarctica.

Each approach we have described has its own strengths and weaknesses, and we need an appropriate mix of tools for both environmental and health protection. While Chapter Three looks in more detail at economic approaches, here we are more concerned with those tools and approaches that relate more directly to technical and technological changes.

## Summary

The scale of issues in urban areas is daunting but, like many environmental issues, the overall effect is often the result of many smaller contributions. Urban populations are growing considerably with implications for resource use and waste generation. In addition there are impacts on the natural environment as well as socio-economic implications. The way in which we perceive these environmental risks and uncertainties for the future may be influenced by many factors including the potential for causing harm and the timescale of effects. Equally important is the need to balance the risks against the benefit of urban development. We have introduced some tools that we will now develop further.

# 3   Resource management

As noted at the start of Section 2, we need to focus on an appropriate scale, so let's start with something with which we are all familiar and can easily deal: a single building – perhaps your home or workplace. We will explore some issues relating to the use of resources and the generation of wastes at that scale. From there we will widen our focus into the urban environment. Diagrams can help us visualize the complexity of the interactions in an urban environment; interactions such as in the flow of resources through the life cycle of something from its design to disposal. So, we will also look at a useful tool: diagramming.

Throughout its life cycle, any building affects the environment. Although site development and construction is a temporary process, it can disrupt the local ecology, and the procurement and manufacture of materials affects other environments away from the construction area. In addition, long-term impacts occur once the building is complete. For instance, the use of energy and water produce air pollution and sewage, and extracting, refining and transporting resources throughout the building life also have numerous environmental effects.

### Activity 2.3

Consider energy use in buildings. Look at Table 2.2 and identify, by function, the major use of energy throughout all sectors in the UK. Can you see any important function that is not listed?

Table 2.2    Energy use for different functions in the UK in 1998

| | Percentage energy used for different purposes | | | Total energy use |
|---|---|---|---|---|
| | **Domestic** | **Services** | **Industry** | |
| space heating | 57.0 | 56.0 | 10.5 | 41.0 |
| water heating | 25.0 | 9.7 | - | 13.5 |
| cooking/catering | 5.2 | 9.2 | - | 4.3 |
| lighting and appliances | 12.8 | 15.9 | 1.0 | 9.5 |
| industrial process use | - | - | 58.5 | 17.2 |
| motors and drives | - | - | 8.2 | 2.4 |
| drying/separation | - | - | 7.1 | 2.1 |
| Other, non-transport | - | 9.2 | 14.6 | 6.2 |
| **Total** | **46.1** | **20.7** | **29.4** | **96.2** |

### Comment

From the table it appears that a lot of energy use in the UK is for heating buildings. In warmer climates, air conditioning may be a major energy use. As you see in Table 2.2, much energy heats houses, but there is also substantial consumption in commercial, public and industrial buildings, all of which form our urban environment.

Did you spot anything missing from the table? You should always look closely at data to see if there are any anomalies or omissions. There is a clue in the last row: transport energy use is missing. In terms of consumption, it about equalled the total energy consumption for the service and industrial sectors combined. We consider transport further in Section 6.3.

○    What happens to the energy that we put into buildings?

●    It makes buildings more comfortable to occupy. However, some of this energy is lost to draughts and poorly insulated structures (see Figure 2.9). These losses help make urban areas warmer than their surroundings – the 'urban heat island' effect – more of that in Section 6.3.

Figure 2.9 shows output flows from the system we are examining: a building. By making the width of the arrows proportional to the magnitude of the flows, their relative proportions are immediately apparent. This approach is known as a 'Sankey diagram' and you will see more examples of this type of diagram later. Information in a diagram like this allows us to identify priorities. The most efficient heating system cannot easily heat a building with draughty windows or door frames, for example. Proper insulation and weather-stripping, combined with an efficient heating system, can provide a warm, comfortable home using minimum energy and minimizing the release of greenhouse gases and other pollutants.

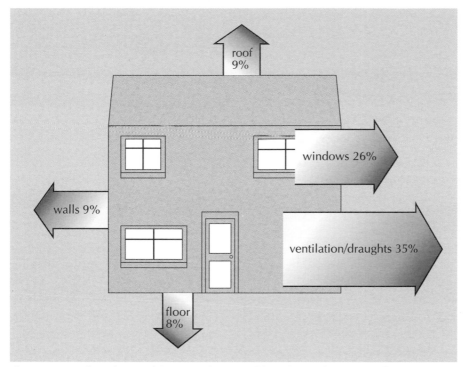

**Figure 2.9** Sankey diagram showing the typical heat losses from a building.

# 3.1  Representing a complex problem in diagrams

While the Sankey diagram may be unfamiliar other diagrams are more recognizable, for example, a route map for a journey or the plumbing and wiring diagrams for a central heating system are examples. Both follow conventions appropriate to the users and aim to simplify complexity. Diagrams are useful because:

See also **Morris et al., 2003** and **Furniss, 2003** on systems.

- activities and processes in real life are often complex, with many interactions;
- diagrams help us understand such complexity by 'mapping' relationships. They indicate what exists and what may be introduced or changed;
- diagrams help us communicate complex situations to others;
- diagrams of complex systems are used in advanced analytical tools, such as life-cycle assessment (more of this in Section 5);
- non-linear systems can be represented.

'System' is a word commonly used in everyday life and we have used it already in this chapter and elsewhere. Write down three or four examples of ways in which you would apply the word to your home life or your work life. It may help to sketch out some components that make up these systems and their relationships. For my list, go to the end of this chapter. Later we will return to the 'systems' concept, but first we will look at some diagrams used by environmental engineers to represent complex systems.

Several types of diagram are used to represent complexity (for example, block diagrams of ecosystems in **Morris and Turner, 2003**). Here are some general diagramming principles:

- Diagramming is not difficult and improves with practice, but care is needed in choosing the type of diagram.
- Diagrams are made up of pictures, words and symbols, often linked by arrows. Words serve as a guide.
- We can distinguish between maps, which can show relationships in space or time, and logical diagrams, which show how things depend on each other. Combining the two may confuse.
- Diagrams can be integrative and can often give a simplified impression of a complex situation better than text.
- Understanding of a situation often grows as it is expressed as a diagram; if not, areas of ignorance are revealed. The diagram also helps communication and is important in modelling and design.
- No artistic skill is needed, but you may need several attempts to produce a good layout. There is no 'right' diagram for a problem, as long as it works for you and other viewers.

Sankey diagrams, graphs and charts are all special types of diagrams that follow conventions and aim to communicate information more effectively than a mass of numbers in a table. As you continue reading this chapter, you will see features of appropriate diagramming techniques.

## 3.2 Input–output diagrams

Figure 2.9 shows outputs from a system, however, we could also show inputs. Figure 2.10 gives a simple representation of a conventional wet central heating system at home. A central block represents the system being considered, with arrows representing the inputs and outputs. Notice that we have drawn a box system boundary  around the whole diagram; we call this the **system boundary**. It defines the scope of our consideration, and indicates that we are considering only the boiler system, and not the whole central heating system throughout the building.

While crude, a diagram such as this can often illustrate what is happening more effectively than lists of inputs and outputs. To demonstrate this, answer the following Activity.

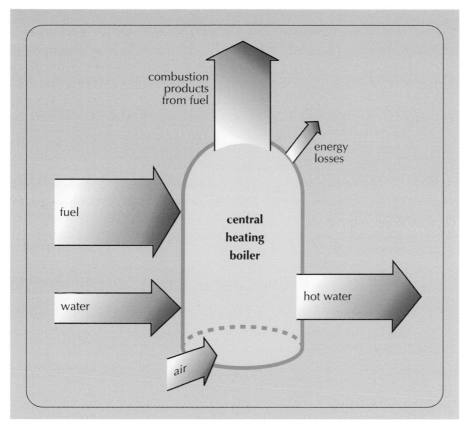

**Figure 2.10** An input–output Sankey diagram of a heating system.

**Activity 2.5**

Make a list of the inputs and outputs to your home like, for example, water. For an answer, go to the end of this chapter.

---

Any resources entering a building system will eventually come out of it. This is an inevitable result of the physical laws of conservation of matter and energy. These state that within any system we should be able to account for the flows of energy and materials into and out of that system (**Reddish, 2003**). For a given resource, its forms before entry to a building and after exit may be different. Many mechanical processes, or human interventions on resources, during their use cause transformations. Inputs to, and outputs from, the building system are diverse, with various forms, volumes and environmental implications.

Ideally we want efficient resource use so that, by reducing waste, environmental risks from pollution, resource depletion, and so on, are also reduced. Strategies for conserving energy, water and other material all have similarities in that they involve maximizing utility and minimizing waste. We can look at energy conservation, followed by water conservation, in more detail.

## 3.3 Energy conservation

After construction a building requires energy throughout its life and the energy consumed for heating, cooling, lighting and equipment operation cannot always be recovered for any useful purpose (see Table 2.3). The environmental impacts of energy consumption by buildings occur primarily away from the building site, through mining or harvesting energy sources and generating power. The type, location and magnitude of environmental impacts depend on the type of energy delivered. Coal-fired electric power plants emit pollutants such as $SO_2$, $CO_2$ and NOx into the atmosphere. Look back at Figure 2.3 to see how UK household emissions of carbon dioxide declined by a quarter between 1970 and 1999. This was primarily because of the switch from coal to gas in homes for heating and by power stations. There is limited scope for a continued decline as around three-quarters of homes now use gas for heating. Alternative energy sources may have problems. Hydroelectric power plants disturb river ecosystems and habitats for animals and plants. Nuclear power plants produce radioactive wastes, with their own special problems (see **Blowers and Elliot, 2003** and **Reddish, 2003**, and Chapter One in this book).

## Energy conversions

One definition of energy is the capacity for doing work. Central to any discussion of energy is the conversion from one form into another. All of these conversions are subject to the First Law of Thermodynamics, which is based on the simple concept that energy can be neither created nor destroyed, but only converted from one form to another. Here, our interest is restricted to indicating the efficiency with which energy conversions take place. Table 2.3 shows a range of machines grouped according to conversion route and the maximum practical efficiency with which they convert energy from one form to another.

Why then are the efficiencies of the machines identified in Table 2.3 not 100 per cent? Technology is purposive: we measure efficiency as the useful energy produced by a machine as a fraction of the total energy input. The Second Law of Thermodynamics tells us that there will be 'losses' but, from the First Law of Thermodynamics, energy is not really lost, merely converted to forms we cannot use. Stating that a large industrial steam boiler has an efficiency of 90 per cent means that 90 per cent of the chemical energy of the fuel used to fire the boiler is converted to useful thermal energy in steam. The remaining 10 per cent of the input energy is dissipated as heat loss up the chimney and through the boiler casing, as Figure 2.10 implies. If we use an industrial steam boiler to drive a steam turbine, we only get 45 per cent of the energy in the steam, which, in turn, only constitutes 90 per cent of the original energy put into the industrial boiler. The overall efficiency becomes 90 per cent $\times$ 45 per cent = 40.5 per cent.

Table 2.3   Efficiencies of energy converters

| Machine | Efficiency (%) | Conversion route |
|---|---|---|
| condensing gas boiler | 88-92 | chemical → thermal |
| industrial boiler – steam | 90 | |
| industrial boiler – hot water | 78-80 | |
| hydroelectric turbine | 90 | mechanical → electrical |
| large electric generator | 90 | |
| small electric generator | 70 | |
| steam turbine | 45 | thermal → mechanical |
| diesel engine | 40 | |
| car engine | 25 | |

## Activity 2.6

An electricity generating station comprises a steam boiler, a steam turbine and a large electric generator. Use the data in Table 2.3 to calculate the overall power station efficiency for electricity generation from a primary fuel such as coal or oil. Look back at the previous paragraphs on energy conversion for a clue on how to do this. For the answer, go to the end of this chapter.

Activity 2.6 illustrates a general point that low overall energy conversion efficiencies can result from a process involving several conversion steps, particularly when there is a transition from thermal to mechanical energy **(Reddish, 2003)**.

Conventional large-scale electricity generation can only be effected with large energy conversion losses: typically, it takes three or four units of primary fuel to yield one unit of electricity. This is of considerable practical significance. We are prepared to pay this 'energy penalty' (which is inevitably reflected in the financial cost of electricity) because electricity is so convenient and capable of efficient end-use. In many applications, such as lighting, electrical machinery and electrochemical processes, it is indispensable, but the environmental impact of the 'penalty' demands that we use electricity efficiently in individual buildings and businesses.

We can 'capture' some of these losses and we are becoming more innovative in this. For example we can recycle the heat where it can usefully be used, or integrate it into power generating systems such as combined heat and power (CHP, see Figure 2.11). These can more than double the overall efficiency of energy use, and small local CHP plants can feed heat to local district heating networks. We can represent these energy flows with Sankey diagrams (Figure 2.12, recall that we introduced these in Figure 2.9).

Combined heat and power (or co-generation) describes the joint production of heat (often as steam) and power (as electricity).

While citywide CHP/district heating systems have attractions, smaller CHP units meeting local heat loads are often preferable. These CHP and district heating

**Figure 2.11** A combined heat and power (CHP) station.

schemes may be attractive if natural biomass is used as a fuel. Cities may not seem an obvious source of biomass, but most generate a surprising amount of wood wastes from parks and other areas. District heating schemes need local users of the energy, and this leads to specific urban design concepts that we will come across towards the end of this chapter.

Returning to buildings, inefficient heating systems can increase the energy losses shown in Figure 2.10 considerably. However, efficiency can be improved, thereby cutting the emissions that contribute to climate change. One option is to replace an existing boiler with a new energy-efficient model. This could also be of smaller capacity than the existing system, especially if draught proofing and insulation have improved the energy performance of the building. Some changes may be costly initially, but pay back in the longer term. There are a number of inexpensive options.

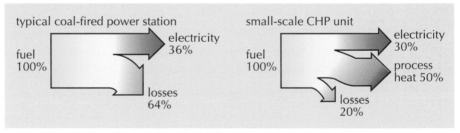

**Figure 2.12** Sankey diagram for electricity generation using a conventional route (power only) and a CHP route in which some losses are used for a process.

A thermostat in its simplest form is a temperature-sensitive switch that controls a boiler or air conditioner. When the indoor air temperature falls below a thermostat setting, the switch moves to the 'on' position. When the air temperature rises to the desired level, the switch turns off the boiler. An air conditioner's thermostat is activated when the indoor temperature rises above a pre-set point and is switched off when the room cools to the desired level. The aim is to maintain comfortable conditions within the room. Lower temperatures are acceptable when working or exercising (or when a room is unoccupied) than when relaxing. So a no-cost way of reducing energy consumption is to set the thermostat to a lower setting at night or when away for long periods, thereby saving energy and money. Although a thermostat can be lowered manually, a low-cost 'technical' option adjusts the setting automatically according to a pre-set schedule. By maintaining the highest temperature for just four to five hours a day instead of 16 hours, programmable thermostats can pay for themselves in four years or less.

# Diagrams with feedback loops

So far we have indicated 'straight through' diagrams that could apply to a process or activity. However, where material or information is generated and is fed back to modify one or more inputs to the system, we can introduce loops into a diagram. So, we could modify Figure 2.10 to show a control loop for water temperature linked to a thermostatic controller on the boiler. We could also show water recirculation through the heating system (Figure 2.13).

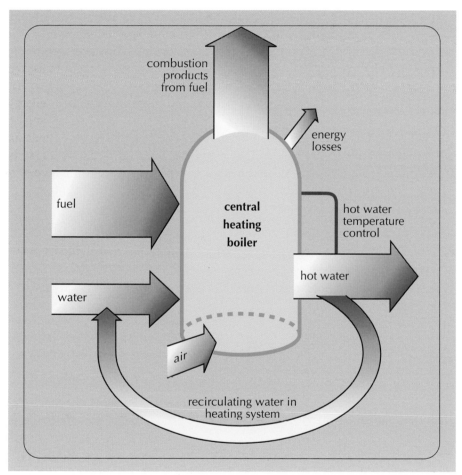

**Figure 2.13** An input–output Sankey diagram also showing 'feedback' and 'recirculation'. The black line around the image denotes the system boundary.

Depending on the heating system design we may have to widen the system boundary. In doing so we may widen the opportunities for energy efficiency available to us, perhaps moving towards a more sustainable future from renewable energy. Many Japanese houses already have solar hot water systems, for example. Micro-CHPs could work at the domestic level, running on gas or even better on renewably-generated hydrogen, with every house being a power station exporting its excess power to the national network.

# 3.4 Water conservation

See also **Brandon and Smith (2003)** and **Furniss (2003)** for water-related environmental issues.

The issues surrounding the use of water in urban areas deserve special attention because both the use of water and the energy for heating or pumping water contribute to resource depletion. Figure 2.3 demonstrated that householders in the UK have been using increasing quantities of water (for drinking, cooking, washing and cleaning, flushing toilets, irrigating plants, etc.). All of this water requires treatment and delivery, which consume energy. Water leaving the building as sewage must also be treated. As we mentioned in Section 2, many people living in cities lack these facilities.

If you look back at Table 2.2 you will see that water heating is the largest domestic energy use after space heating. However, as Box 2.3 demonstrates, there are many ways to improve the efficiency of water use.

## Box 2.3  Improving water management at home

Even if you use a shower rather than a water and energy-inefficient bath, replacing a standard showerhead (with a flow rate of 14.3 litres/minute) with an energy efficient showerhead (flow rate less than 11 litres/minute) can conserve water effectively.

One quarter of your hot water use is for laundry. Rinse in cold water – it does not affect the cleanliness of the wash. Improve the energy efficiency of your hot water tank with an insulating blanket and pipe insulation. An insulating blanket around the tank can reduce energy use by 10 per cent and cut $CO_2$ emissions.

Conventional bathroom and kitchen taps can use at least 12 litres of water per minute. Higher efficiency flow controls require much less water. Commercial tap flow controls use up to 85 per cent less water.

### Activity 2.7

How much water per year will a low-flow showerhead save in a household that altogether takes 16 minutes worth of showers each day? For the answer, go to the end of this chapter.

### Activity 2.8

How may toilets with a low-volume flush contribute to resource efficiency? For the answer, go to the end of this chapter.

## Summary

So far we have seen the scale of the problem of urban growth and some implications of resource use. Uses of energy and water are especially important issues that can benefit from efficient management, starting with simple designs that avoid or reduce use, and only then progressing to more complex control systems. Applying this hierarchy is one 'tool' that environmental managers can use to conserve resources and reduce the risks of environmental harm. Let's develop these ideas further, continuing, at first, with the example of water.

# 4   Integrated resource management: relationship diagrams and systems

The water resource management system is one complex example of our urban infrastructure (as mentioned in the answer to Activity 2.4). To supply clean water we need sources of water, treatment facilities to make it suitable for use and a distribution network. Similarly, to deal with waste water we need a collection network, treatment facilities and a means of disposing of treated water and residues (Figure 2.14). We can show these relationships diagrammatically (Figure 2.15).

Figure 2.14  Sewage treatment works

Such a system is made up of components that may themselves be sub-systems, but how far we break down a system depends on our viewpoint of a problem. An architect designing a new building will consider the boiler as a system component, whereas an engineer seeking the cause of excessive energy consumption will consider the boiler as a system and the heat exchanger and burner as system components.

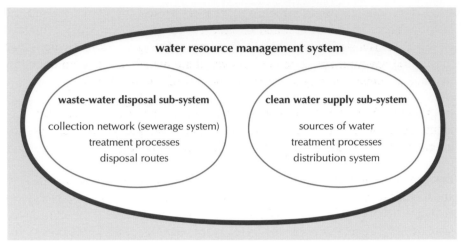

Figure 2.15 An urban water resource management system.

We have seen that when we define a system or sub-system, certain things are part of the system and others are not. We draw a boundary around certain components, and defining the boundary is useful for clarifying the concept of the system. The boundary relates to what a system does or produces, and we reflect this in the name of the system (as we saw in the water resource management system). Components of a system are necessary for it to function, but other factors can be outside the boundary, where they are not essential but nevertheless may influence the system. Whether people are inside or outside the system under consideration is important to how they view it. For environmental issues, people should be *inside*, not outside, the system boundary, since our perceptions and actions are influential.

## Activity 2.9

Describe the functioning of a motor car in terms of systems. For example, the fuel system and the ignition system work together, but there are many others you can consider depending on where you draw your boundary. For an answer, go to the end of this chapter.

Just as there may be a lower limit to how far we define a sub-system, so there is a wider limit. This is often placed where a system can be defined practically without too much complexity. This boundary may be at the household, at the company or at local or national government level, depending on the issue. For simplicity, we began our analysis at the manageable level of the household.

## Activity 2.10

Represent your home in terms of sub-systems and the wider environment in which it operates. For example, we have already looked at the heating system inside a home. Outside is the transport system, the local government system and so on. Now build on these examples.

In addition to its commonly understood meaning, we use the word 'environment' in this chapter to describe much of what exists outside a system boundary. Clearly there is a similarity, but the environment of a system involves more than the natural environment. It comprises everything outside the system boundary that may influence directly or indirectly the activities of the system.

In answering Activity 2.10 and analysing your own home system, you should have identified sub-systems such as heating, electrical, plumbing etc, and the interactions with other systems in the wider environment.

Environmental influences can be diverse, but fall into social, political, economic and technical domains as well as natural factors. For example, within the market economy (**Castree, 2003** and Chapter 3 in this book), the damp climate and the fast-flowing streams to provide waterpower were technical influences that encouraged the establishment of the cotton industry in Lancashire during the industrial revolution. Current influences on businesses might include the range of financial pressures that contribute to the economic environment, legislation that reflects the political environment, and so on.

# 4.1  Identifying and solving problems

Facing up to the pressures of a changing culture and the demands of society for greater environmental care is challenging. We can deal with the risks of hazardous industrial pollutants by achieving cleaner production or 'eco-efficiency' using some of the principles we have seen already. This goal can be partly achieved by industry reducing its raw material inputs – chemicals, natural resources, energy, and water – and at the same time reducing air, water, and solid pollutants for each unit of production. This push toward cleaner production is typically driven by environmental and economic concerns, although it seems certain that cleaner production would also benefit public health.

The solutions to environmental problems must not ignore the role of people, and cannot be done in one step. One toolkit for tackling problems might involve a three-stage linear process:

- Diagnosis – adopt an approach to the problem and describe it as a system from the approach adopted, in order to devise appropriate objectives and measures.
- Design – propose and explore alternative solutions.
- Implementation – evaluate and implement options as appropriate.

### Activity 2.11

From your experience or knowledge of problems of the type described, suggest any shortcomings to this linear approach.

When you read the word 'linear' in Activity 2.11 you may have realized that dealing with a problem of this type is not a linear process. For example, when you decide to move house, you face a complex problem: you consider the 'environment' of different areas, the economic aspects, the different 'hard' features of the house itself, and effectively go through 'diagnosis' and 'design' phases several times. The problem-solving process is iterative. You go through a cyclical process in which your understanding improves and options are checked against objectives – effectively feedback. Those in charge of environmental management systems use a similar approach. An organization identifies its most important priorities taking account of its business environment, sets objectives, implements solutions and then reviews progress before going round the cycle again with the aim of continual improvement.

If we look back through industrial history, we see another way in which there has been a change from a linear approach in dealing with waste. Linear industrialism, as it evolved since the late eighteenth century, followed the sequence found in Figure 2.16.

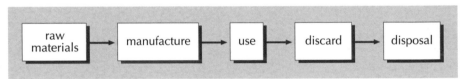

Figure 2.16 The linear approach to dealing with waste.

By reviewing whether discarded resources are in fact waste, companies discover that some may be diverted for re-use (see Figure 2.17).

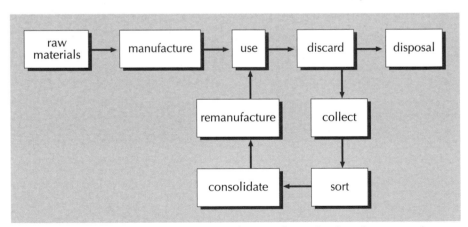

Figure 2.17 Our approaches to dealing with waste have developed a re-use phase.

## 4.2  Resource conservation

Recall that, earlier in this Chapter, I suggested that resources often change during their use. We call goods that have lost their original usefulness 'waste', but waste can be a resource for another use. It is important to question what is waste, and if it is inevitable.

Materials efficiency and waste prevention require a cyclical approach rather than typical linear 'extract, use and discard' approach to manufacturing and using resources. Figure 2.18, which shows resources used to provide domestic services, includes water inputs that may be abstracted from a river, used, discharged for treatment, returned to a river and abstracted for use again many times during the flow down a river system. However, 'recycling' is not the only way of increasing the likelihood of resource conservation. The '3Rs' – reduce, re-use, recycle – provide a hierarchy to change our current resource use patterns and so can shape our future use. These principles offer an approach to managing environmental problems of pollution and resource depletion.

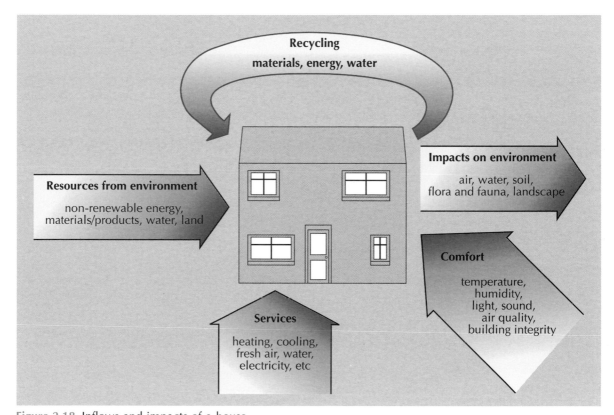

Figure 2.18 Inflows and impacts of a house.

## Reduce

'Reduce' or 'source reduction', is the first step in managing efficiency and preventing waste as we saw earlier in the chapter for energy and water. We can do this by:

- increasing the efficiency of the use of energy, water, paper, cardboard, glass, metal, plastic and other materials;

- reducing the use of non-recyclable materials;
- replacing disposable materials and products with reusable materials and products;
- reducing packaging;
- eliminating or reducing the amount or toxicity of materials before they enter the waste stream.

The replacement of ozone depleting substances (such as CFCs) with substitutes is an example of eliminating harmful materials.

The potential for source reduction can be summed up in three key words:

- time – is equipment in operation when it could be turned off? Ideally this should be done automatically.
- level – does equipment operate at too high a level, perhaps pressure, flow or temperature (or too low a temperature for refrigeration)?
- efficiency – are resources used to the best advantage at all times?

Examples illustrating these principles are:

*Lighting* – is it on when not required, too bright or provided by inefficient filament lightbulbs?

*Boiler* – is it on standby duty when it could be switched off, or operating at a higher pressure than necessary and with low combustion performance?

*Water* – do taps/hoses operate at too high a flow or for too long, and are there leaks?

*Heated storage tanks* – are they heated unnecessarily, are contents at too high a temperature, are outer surfaces uninsulated?

*Computers* – are monitors on energy saving mode and computers off when not needed?

*Television* – a television on standby uses significant energy.

These principles of time, level and efficiency are just 'good housekeeping'. Such better practices and procedures can improve environmental performance at little cost with considerable environmental and economic benefits.

### Activity 2.12

Complete the table below using the key words 'time', 'level' and 'efficiency'. One example has been given to help you.

---

**Practical examples of good housekeeping**

---

adjust boiler controls to give optimum combustion performance

ensure that hot water storage temperature is not higher than necessary

improve leak maintenance

isolate unnecessary sections of heating system to empty rooms

keep close control on space heating level and times                           time and level

switch off unnecessary lights

replace defective lights by high efficiency tubes

ensure vehicle is correctly maintained

control/limit electrical space heating

---

# Re-use

'Re-use' is the next step. Effective re-use preserves the structure of a material or article, as with the returnable glass milk bottle and similar beverage containers. Other examples are intact or repairable home appliances, industrial appliances, household furniture, clothing, engines and gearboxes (see Figure 2.19), intact materials in demolition debris and building materials.

# Recycling

'Recycling' is the third step, involving converting manufactured articles back into raw material for remanufacture. By replacing virgin materials with recycled feedstock, we preserve natural resources and energy. As building users, we can encourage recycling by providing facilities for separating wastes for recycling. This is increasingly done domestically in many countries, and is also done in business.

**Figure 2.19** Gearboxes at a vehicle-dismantling centre may be re-used.

By recycling we help reduce waste disposal to landfill and future methane emissions (typically 66 kg of methane are emitted per tonne of waste) (see Box 2.4). Methane is a major contributor to climate change (**Blackmore and Barratt, 2003**). If you have a garden, recycling waste through composting can reduce the need for chemical fertilizers whose manufacture and use also emit greenhouse gases. Nitrogen fertilizer use, for example, accounts for significant emissions of nitrous oxide ($N_2O$), a potent greenhouse gas. Chemical fertilizers also add to water pollution problems.

We could add a fourth 'R' – 'rebuying'. Without markets for recycled goods, the cycle is broken. We need encouragement to buy products made from recycled materials and a change in the perception of the public that re-used or recycled products are inferior.

## Box 2.4  Design for recycling

A trend since the 1990s has been to design products for recycling at the end of their service lifetime. Computers and other products are increasingly designed with recycling in mind, but a prime example can be found in the automobile industry. A modern car is made up of many different components, each fulfilling a specific function and linked together in various ways. A typical composition is:

| Material | Percentage of total weight |
|----------|----------------------------|
| Metal    | 73                         |
| Plastics | 9                          |
| Rubber   | 4                          |
| Glass    | 3                          |
| Others   | 11                         |

So, several recycling loops are needed in order to recycle a scrap car.

Figure 2.20  End-of-life vehicles are not necessarily waste.

A small family saloon weighs approximately 1,000 kg and its average life is about 10 years. With something over 20 million new registrations annually in Europe alone, that represents 20 million tonnes of component materials available for recycling every year. Approximately 98 per cent of ferrous metal and 90 per cent of non-ferrous metal is currently recycled.

Recycling depends on the last owner of a vehicle handing it over to a specialist firm equipped to recognize the primary material used for the individual components in that model. Ideally, recycling follows the hierarchy, with re-use being preferred to recycling. So, during a vehicle's lifetime, certain components, such as engines and gearboxes, might be replaced with reconditioned components and worn-out parts remanufactured to the same standards as new ones (see Figure 2.19).

Only if the precise specification of components is known can they be 'recycled' at the highest level, that is new plastic fuel tanks from old ones, new bumpers from old, and so on. New components with exact technical specifications can only be made directly from recycled material of the same quality. If the quality is lower, the material must be downgraded in the recycling loop, if it can join it at all.

The main aim, then, is to recycle at the original level. If that is not possible, downgrading to the next quality level might be feasible, for example to make wheel arch shells or internal compartments. The penultimate resort, at least for combustible materials, is to burn them to recover some of the energy that was used in their production. Landfill is the last resort for non-combustible material but will become increasingly costly over the coming years through taxation. Taxation will provide an extra impetus to recycle if at all possible.

In future, at the design stage, records will have to be kept and individual parts marked with a code identifying the materials used; many plastics already have embossed identification marks (Figure 2.21). When it is no longer worthwhile fitting reconditioned components, products will be discarded and disassembled, and the various parts sorted by material type (identified by the marked code). What remains can be shredded and mechanical sorting used to separate materials for transport to their primary producers in the appropriate metallurgical and chemical industries.

In addition to end-of-life vehicles, the flow of electrical and electronic equipment is one of the fastest growing waste streams in the European Union, constituting four per cent of the municipal waste today. This level increases by 16 to 28 per cent every five years – three times as fast as the growth of average municipal waste. Furthermore, electrical and electronic equipment is one of the largest known sources of heavy metals and organic pollutants in municipal waste. Bearing in mind the resource-intensive production of this equipment, the legal requirement to recycle it should lead to significant resource savings. This is just one example of how legal pressures are driving the reduction in wasteful consumption of natural resources and the prevention of pollution.

## Activity 2.13

Look at the labels on some plastic products around you to confirm which products are used for which purpose.

PETE

PETE is an acronym used by manufacturers to mark and identify plastic bottles or containers made from polyethylene terephthalate for the purpose of recycling. The acronym PET is more generally used within the chemical industry for polyethylene terephthalate. PET is the acronym accepted by standards organizations.

PET, also known as polyester, is a popular packaging material for food and non-food products because it is inexpensive, lightweight, resealable, shatter-resistant and recyclable. It is clear and has good moisture and gas barrier properties.

HDPE

Bottles made from High Density Polyethylene (HDPE) come in both unpigmented and pigmented resins. The former is translucent, with good stiffness and barrier properties. So, it is ideal for packaging products with a short shelf life such as milk. Good chemical resistance allows it to be used in containers holding household or industrial chemicals.

PE - HD

Plastic bag manufacturers use a different symbol for HDPE as shown to the left:

V

Vinyl, or polyvinylchloride (PVC) has stable electrical and physical properties. It has excellent chemical resistance and good weatherability. Its flow characteristics make it ideal for injection moulding.

LDPE

Its toughness, flexibility, and transparency, make low density polyethylene (LDPE) ideal in applications where heat sealing is necessary. It is also widely used in wire and cable insulation.

LD-PE

Plastic bag manufacturers use a different symbol for LDPE as shown to the left:

PP

Polypropylene (PP) has the lowest density of the resins used in packaging. It is strong and resistant to chemicals. With a high melting point it can be used where a container is filled with a hot liquid.

PS

Polystyrene (PS) can be made into rigid or foamed products. It has a relatively low melting point.

OTHER

The 'Other' category includes any resin not specifically numbered one to six, or combinations of one or more of these resins.

**Figure 2.21** Identification codes aid plastic recycling.

When waste is truly waste, one way of recovering the energy in the waste is to collect methane generated in landfill sites as the organic matter degrades and to burn the methane as a fuel (Figure 2.22). This is rarely popular, given public fears of the risks from landfill sites, and there are limits to the number of sites we can accommodate near cities.

Figure 2.22 Recovery of methane at a landfill site.

Another option is incineration. Like all combustion processes, incineration inevitably generates carbon dioxide, which contributes to climate change. However, recovering energy by waste combustion is at least partly greenhouse neutral in that much original material is biological. So, while incineration generates carbon dioxide, vegetation had absorbed carbon dioxide for photosynthesis during growth. However, public concerns about emissions from incinerators are not easily assuaged. While technically the environmental concerns over incineration can be dealt with, public perceptions are less easily resolved (where might 'incineration' fit in Figure 2.7?). A less problematic 'energy-from-waste' process is the established technology of using methane from sewage sludge digestion, and sewage is one resource that cities have in plenty.

## Summary

Building on our previous summary on the role of design and management in resource conservation, we have the sequence:

- reduce, or waste minimization at source;
- re-use;
- recycle;
- incineration with energy recovery;
- incineration without energy recovery;
- landfill.

This is known as the 'waste management hierarchy'. It places alternative waste treatment options in a fixed order of preference with waste minimization at source as the most environmentally preferred (least environmental impact) option and disposal to landfill as the least environmentally preferred (most environmental impact) option. Design for disassembly and recycling is an important tool to help us improve our future use of resources and so addresses our key questions on managing environmental problems and providing opportunities for sustainable development.

While this hierarchy is useful, don't regard it as immutable and be careful before drawing conclusions on the most environmentally preferred practice. So, making packages less heavy may not be a good idea if there is greater waste through breakage/damage of the products packaged within it.

Whether or not you view it as strictly renewable, the use of domestic and commercial wastes as a fuel is commercially attractive. It becomes even more interesting if it is linked with co-generation – that is using the heat produced as well as the electricity, in combined heat and power (CHP) plants. We can also apply our hierarchy to buildings.

# 5    Building blocks: life-cycle thinking

So far we have looked at design and environmental issues at the level of an individual building. If you will excuse the pun, buildings are the building blocks of our urban areas and their design, construction and maintenance has a tremendous impact on our environment and our natural resources. Look back at the Sankey diagram that represented a single dwelling in Figure 2.9. Attached houses and flats can 'capture' some of the heat lost from neighbouring units. So, in order to understand their effects, we need to widen our system boundary over time and space. Doing so helps to identify where we can make improvements to urban environments.

## 5.1 Flow diagrams

The input–output diagrams that we considered earlier can be applied to the manufacturing processes that produce the products we use in our homes and businesses, and even to construction of the building itself. However, flow diagrams may be more useful to represent increasingly complex flows of materials or energy over space and time. Materials flow diagrams use nouns (names of equipment or departments) to describe stages in the transformation process. Process flow diagrams are more general, with boxes being labelled with the name of the process being carried out. For example, Figure 2.23 shows a process flow diagram for the manufacture of knitted garments – perhaps one of the inputs (clothing) that you identified in Activity 2.5.

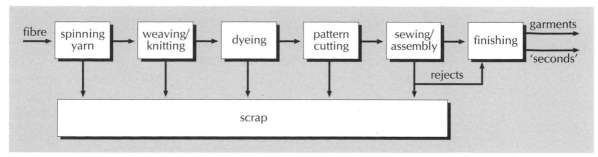

Figure 2.23 Process flow diagram representing garment manufacture.

Flow diagrams may help widen your thinking if you are seeking possible problem areas, alternative methods for carrying out a series of tasks or to visualize inter-relationships, such as where there are uses for waste, as shown in Figure 2.23. The manufacture of such products is perhaps easy to visualize as a process with inputs and outputs, but what about non-manufacturing processes carried out by the large sector of the economy involved in providing services? The 'products' of education, health care, travel, entertainment industries and the financial sector may be less tangible, but nevertheless require inputs, produce outputs and generate waste. All are part of our urban systems.

The conventional view of building construction is that it involves a linear process of design, construction, operation and maintenance and demolition (see Figure 2.24).

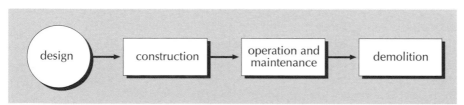

Figure 2.24 A conventional view of a building life cycle.

Consider for a moment what information Figure 2.24 gives you. Very little you may conclude. It ignores environmental issues (related to the procurement and manufacturing of building materials) and waste management (re-use and recycling of architectural resources). Analysing the typical life cycle of a building starts with the building materials from extraction of natural resources through each stage or process where the materials are modified and transported for use in construction and operation of the building, and ends with its final demolition. Figure 2.25 shows a typical flow. Notice that this reflects activities through the life of a building. If we have two optional life cycles with different elements within them, comparing options in this way can help us choose between them.

The time line shown at the top of Figure 2.25 represents the relative length of stages in the life cycle of the building. The resources include such items as iron ore, minerals, rock, petroleum, trees and other vegetation, soil, water, and so on –

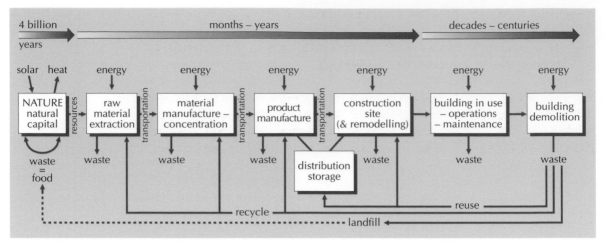

**Figure 2.25** A more detailed building life cycle showing recycling, reuse and landfill.

anything that is used in the construction and maintenance of the building. These materials are processed into a form that makes them suitable for the manufacturing of a building component, for example, the processing of iron ore into steel, petroleum into nylon, and trees into timber. This refined material is next used to make products, for example, steel beams, nylon carpets, timber floors or furniture. Finally, products are used in the construction and operation of the completed building.

Each stage requires energy to add 'value' and create a higher order of product quality, and each stage generates waste. Only at the first stage, the natural capital stage, is the energy input entirely solar and the waste output totally recycled over geological time. All other stages have human involvement, with various energy sources and with waste being re-used, recycled or released into the air, water or ground. In principle, re-use and recycling could occur at every phase, but is only illustrated for the last phase in order to keep the diagram simple. Notice that we also omitted design from Figure 2.25. However, design is clearly important to the development of sustainable opportunities in that we can design for both the efficient use of resources and for future disassembly and recycling.

Many building materials are brought onto building sites, primarily during the construction stage. The waste generated in construction and installation is significant. A building that is oversized for its design purpose, or has oversized elements, will consume materials excessively. Waste prevention begins upstream with design and the development of material-efficient building technologies. For example, by trimming materials to fit non-modular spaces we generate excessive waste, so design that matches the sizes of construction boards, for example, can help reduce waste.

Recall that an effective method for material conservation is through the use of existing resources. Most buildings outlive the purpose for which they were designed but many of these buildings can be converted to new uses at a lower cost than new construction. This is an important example of re-use – high up in our hierarchy. In addition to conserving materials, if a building is adapted to new uses, the energy associated with its materials and construction is conserved. This energy is considerable. It includes not only the sum of energy in the materials, but also the energy used in constructing the building.

When buildings have to be demolished they can provide resources for new buildings. Some, like architectural features, bricks or windows, can be re-used whole in the new structure. Other building materials, such as wood, steel, and glass, are easily recycled. Again we see our hierarchy applied. So we have seen some further application of the approaches to managing environmental problems. Now we turn to the issue of choice between environmental options.

## 5.2 Life-cycle assessment

Recall that we asked you to list the inputs to your home in Activity 2.5. We could expand this list of materials to specify individual items. We could also quantify all inputs and outputs. We call this **inventory analysis**, a process that quantifies raw material and energy consumption together with solid wastes, emissions to air and water (the environmental burdens) for all processes within the system boundary. We can use this inventory of environmental burdens in the generation of an ecological footprint (**Morris, 2003**). It is also an important stage in the environmental management tool known as life-cycle assessment. By looking at the whole life cycle of something we can identify opportunities for conserving materials and minimizing environmental impact.

*inventory analysis*

Considering the whole life cycle of a process or product can be useful for comparing environmental implications. Known as **life-cycle assessment** (LCA), this tool has been defined as:

*life-cycle assessment*

> Systematic set of procedures for compiling and examining the inputs and outputs of materials and energy and the associated environmental impacts directly attributable to the functioning of a product of service system throughout its life cycle.
>
> (SETAC, 1993, ISO 14040, Section 2)

LCA provides an indicator of the overall environmental impact of a process or product, incorporating raw material extraction, manufacturing, use, re-use or recycling and final disposal. Two important parts of LCA are the inventory and impact assessment.

We first need to decide on our basis for comparison, using what is known as the 'functional unit'. This can be difficult: consider for a moment whether we can compare the environmental impact of two paints over their life cycle on the basis of volume. We may want to compare organic solvent-based paint (with a strong odour and contributing to ozone pollution by emitting volatile organic compounds (see **Blackmore and Barratt, 2003**)) against a water-based paint (with low odour but may contribute to water pollution).

A comparison of volumes seems logical, but the two paints may not have the same covering power, nor may one last as long as the other. We may have to use more than one coat of one product or repaint more often. So, we cannot compare the two on the basis of volume, but need to include some measure of covering power and longevity.

We also have to define the 'system' to be studied. System boundaries generally include:

- the main production sequence, i.e. extraction of raw materials up to and including final product disposal as in Figure 2.25;
- transport operations;
- production and use of fuels;
- generation of energy;
- disposal of process wastes.

Usually system boundaries exclude:

- manufacture and maintenance of capital equipment;
- services of manufacturing establishments, i.e. heating and lighting;
- factors common to each of the products or processes under consideration.

Once we have set our system boundaries, we can gather the data that will form the basis for our assessment.

Inventory analysis lists inputs and outputs for all stages of the life cycle in terms of energy and materials inputs and emissions to air, water and solid waste (unfortunately, we may have difficulty getting some information, for example, from organizations that don't quantify their wastes). From this inventory, energy inputs for the various stages can be added together, as can the material inputs.

### Activity 2.14

Consider an inventory analysis in a life-cycle assessment adding together emissions to atmosphere such as:

$$2g \text{ of } CO_2 + 3g \text{ of } NO_x = 5g \text{ of total emissions}$$

Does this make sense?

### Comment

Mathematically this may be true, but it is meaningless, because of the different environmental effects – climate change from $CO_2$ while NOx contributes to acidification and tropospheric ozone (**Blackmore and Barratt, 2003**).

Comparing impacts can be problematic. Above, I mentioned that organic solvent waste contributes to ozone pollution, while a water-based paint may cause water pollution: how can we decide which is more important? Similarly, how do we compare different environmental problems associated with carrier bags (see Figure 2.26)? No product is better in every respect.

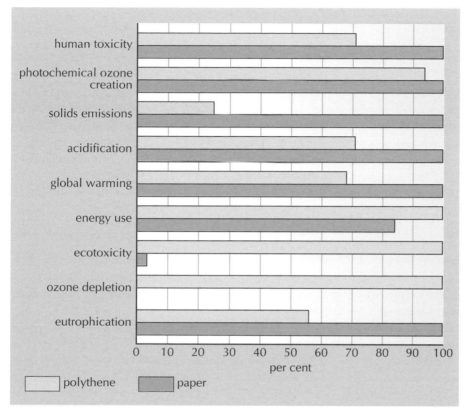

Figure 2.26 LCA comparison of the environmental impact of paper and plastic carrier bags.

We could assign a weighting factor to the impacts, enabling us to calculate a single score, but this involves us making a judgement on the relative importance of, say, climate change and acidification. Such 'valuation methods' are controversial because of the divergent assessments involved.

See **Hinchliffe and Belshaw** **(2003)** and **Burgess (2003)** for different takes on valuation.

## Activity 2.15

LCA is a tool for helping us choose between approaches for managing environmental problems, but it has limitations. Prepare your own summary of the benefits and difficulties of LCA. For my answer, go to the end of this chapter.

Further discussion of LCA is beyond the scope of this chapter. However, you should recognize that LCA is one tool for managing environmental problems. It can help quantify environmental impacts to enable us to make choices between options. Now we will look at what we can do about our urban environmental futures.

## Summary

In our environment we come across many complex flows of materials and energy over time and space. Flow diagrams help us represent these flows and identify areas for improvement, especially by reducing waste. Extending the system boundary to include the whole life cycle of an activity or process is an important approach to help us compare different products or processes. However, there are difficulties in making such life-cycle assessments, including problems in collecting information and making appropriate comparisons of functions. Such difficulties produce large uncertainties.

# 6    Urban development

Having looked in detail at individual buildings, we can now return to the city-scale. How do we design or plan better cities? We look at four possibilities: reducing sprawl, industrial ecosystems, traffic management and urban climate management.

One perspective on urban development considers three phases. First, we aim to meet the basic needs of shelter, water supply, health and nutrition, employment and education. Then our emphasis switches to addressing environmental problems resulting from rising affluence, such as traffic congestion, increasing waste and pollution of all kinds, as well as the need for investments in services such as water supply and sanitation, wastewater treatment, flood control, and urban transportation systems. We also need to consider strategic land use. In the third phase, our attention turns to consolidating urban services to meet high environmental standards, redevelopment of inner city areas, investments in public transportation and so on.

Apart from increasing amounts of wastes in general, the rapid growth of cities has, together with associated industry and transport systems, caused an equally rapid increase in urban air pollution. This comes mainly from fossil fuel combustion in the energy, industrial and transportation sectors. Use of poor quality fuel (eg. coal with high sulphur content), inefficient energy production and use, poor condition of vehicles and traffic congestion all play a role. The way we manage environmental issues such as waste and air pollution depends very much on how we as individuals, and in groups, make conscious decisions to use fewer resources and change behaviour. So if you recall my comments in Section 4, we must be inside our system boundary. The overall impact at the urban or wider level is the aggregate of our individual and group actions.

# 6.1  Urban design

We could define urban sprawl as when the rate at which land is converted to non-agricultural or non-natural uses exceeds the rate of population growth. It may appear as low density development. Sprawl converts farmland, forests, and so on, into impervious surfaces, thereby increasing flooding and runoff, and adding pressure to farm on marginal lands that formerly served as wildlife habitat.

---

**Activity 2.16**

Consider for a few moments some other environmental effects of urban sprawl. Write down your suggestions.

---

Although cities can sometimes create new ecosystems, negative impacts come more easily to mind. Loss of farmland may cause degradation of natural ecosystems, such as remaining woodland and wetlands, and hence pressure on species. More flooding may occur as impermeable surfaces such as roads and parking areas increase. Loss of wetlands that provide flood control can increase the number and intensity of floods. Through increased urban runoff, destruction of wetlands and soil erosion, sprawl threatens the quality of both ground and surface waters (water pollution).

Less obvious effects are that sprawl significantly increases the costs of infrastructure such as sewers, storm water retention and drinking water treatment because of the need to provide services in the suburbs while having to maintain the same infrastructure in the ageing core. Environmental decay occurs in the urban centre leading to brownfield sites, some of which may be contaminated by former industrial activity such as chemical works (see Box 2.5). Not only are there more cars on the roads, but sprawl makes alternative forms of transport uneconomic, forcing us to drive our cars further, increasing congestion and emissions of greenhouse gases and the precursors to ground level ozone (smog).

So, the complex problems caused by urban sprawl can include increasing traffic congestion and commuting times, air pollution, inefficient energy consumption and greater reliance on oil, loss of habitats, inequitable distribution of economic resources and the loss of a sense of community. Remember that in this chapter the focus is on technology. Bringing in social and other dimensions would make this view much more complex.

We have seen that sprawl increases problems for and from cities, especially linked to the impact of goods, services and people travelling around. What if we bring activities closer together, perhaps by 'reusing' land? Can this be a useful way of dealing with some environmental problems?

## Box 2.5  Recycling land

One 'answer' to urban sprawl is to recycle urban land. 'Brownfields' is a term used for old industrial or former industrial sites whose redevelopment is hampered by potential contamination and limited market demand for new uses. Brownfields redevelopment is more complicated than urban economic development in general because environmental, legal, economic and community factors are all involved. Apart from the direct costs of dealing with contamination, there may be uncertainties about regulatory requirements for remediation, and about liability for such work. Such uncertainties have reduced investment confidence and deterred development, contributing to land remaining derelict. Land in these circumstances can have a negative value, and may have been abandoned by its owners. As well as adverse localized social and economic effects, the lack of confidence in brownfield sites arising from contamination and its associated uncertainties can also lead to increased and unwanted pressures for new development on greenfield sites and resulting urban sprawl. Brownfields redevelopment has become a central component of urban redevelopment in older industrial cities. In the UK, for example, the government aims to increase development on brownfield sites, but this raises policy issues about the problems and risks from contamination of the land (Figure 2.27).

**Figure 2.27** A reclaimed landfill site, after being capped, requires long-term monitoring and has restricted use.

# 6.2 Industrial ecosystems

The industrial ecosystem concept illustrates life-cycle thinking and can help combat sprawl. It involves optimizing the use of materials and embedded energy, minimizing waste generation, 'closing loops' by making maximum use of recycled materials in new production and re-evaluating 'wastes' as raw materials for other processes. Such schemes also imply going beyond 'one-dimensional' recycling of a single material or product – (for example, aluminium beverage cans). One model involves complex 'webs' between organizations within an urban area, rather like a natural ecosystem – hence the term 'industrial ecosystem' (see Box 2.6).

## Box 2.6   The industrial ecosystem concept

The ideas outlined above are not new. The first use of the industrial ecosystem concept was at Kalundborg (see Figure 2.28 overleaf). Asnaes, the largest coal-fired electricity generating plant in Denmark (see Figure 2.39 overleaf), began supplying process steam to the Statoil refinery and Novo Nordisk pharmaceutical plant. Around the same time it began supplying surplus heat to a district heating scheme that replaced 3,500 domestic oil-burning heating systems. Gyproc, the wallboard producer, had bought surplus gas from Statoil since the early 1970s, but in 1991 Asnaes began buying all the refinery's remaining surplus gas, saving 30,000 tonnes of coal a year. This initiative was possible because Statoil began removing sulphur from the gas, to make it cleaner burning. Kemira bought the sulphur for making sulphuric acid. Asnaes also desulphurized its airborne emissions, yielding 80,000 tonnes of calcium sulphate as a by-product for sale to Gyproc as 'industrial gypsum' (replacing mined gypsum previously imported). In addition, fly ash from Asnaes is used for cement making and road building.

Asnaes also used its surplus heat for warming its own seawater fish farm, producing 200 tonnes of trout and turbot a year. Sludge from the fish farm served as fertilizer for use by local farmers. Asnaes used remaining surplus heat for a 37-acre horticultural operation under glass. High nutrient-value sludge from fermentation operations is normally regarded as waste, but Novo Nordisk treated it by adding chalk lime and holding it at 90 °C for an hour to neutralize micro-organisms. Then 330,000 tonnes a year served as liquid fertilizer for local farms.

Most of the Kalundborg exchanges were between geographically close participants – in the case of thermal transfer this is clearly important, as infrastructure costs are a factor. However, for other 'wastes', proximity is not essential: the sulphur and fly ash were supplied to buyers further away.

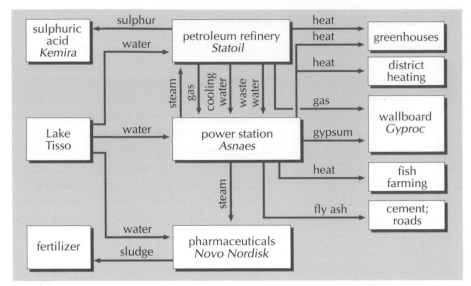

**Figure 2.28** Industrial ecology at Kalundborg.

**Figure 2.29** Asnaes, the largest coal-fired electricity generating plant in Denmark.

# 6.3 Urban traffic management

Imagine living in a city where a haze of smog permeates the streets. Imagine these streets filled with traffic jams and exhaust fumes. You cover your nose and mouth with a handkerchief to avoid exposure to dust and particulate matter in the air but your eyes water and your nose runs. A jogger passes you wearing a facemask. You have in your mind the state of affairs that until the late 1990s existed in Mexico City. Its air was once ranked by the World Health Organization (WHO) as the most contaminated in the world. The health consequences of exposure to polluted air are considerable – many respiratory diseases are associated with air pollution. In addition, urban air pollutants contribute to problems such as acidification and climate change, which have impacts on crop productivity, forest growth, biodiversity, buildings and cultural monuments. Clean air is a basic requirement for human health and welfare as well as for sustainable economic development.

Air pollution is widespread across megacities and especially in cities in developing countries. In most Asian megacities such as Beijing, Delhi and Jakarta air pollution has worsened through the cumulative effects of population growth, industrialization and increased vehicle use. As the urban populations have increased, so has the number of vehicles and so has fossil fuel consumption. Not surprisingly, air pollution also rose dramatically, and health has deteriorated. Children are particularly at risk because their lungs are still developing, they breathe faster than adults, and they tend to absorb pollutants more readily. To various degrees, most cities face similar problems of congestion and pollution from urban traffic (Figure 2.30 overleaf). So, we return to one of this book's three key questions, what tools and approaches are available for managing environmental problems and how useful are they? Different strategies yield various degrees of success, but none is perfect. One approach calls for replacing old automobiles and industrial machines that are highly inefficient, but the substantial costs involved may exceed the financial capacity of many developing countries.

In response to its reputation as one of the world's smoggiest and most unhealthy cities, Mexico City instigated a 10-year air quality plan, which involved measures such as banning higher-emission vehicles from the roads on the frequent days of pollution alerts, retiring old lorries from the road and limiting the hours that all lorries can spend in transit. Other steps included using more natural gas in public buses, giving taxi drivers incentives to buy newer cars and introducing hybrid gas-electric cars and buses.

Earlier measures, such as keeping older and more-polluting cars off the street one day a week, introducing catalytic converters and lowering sulphur levels in petrol, resulted in a decrease in vehicle emissions, blamed for 70 per cent of the city's air pollution. Levels of smog, lead, sulphur dioxide, carbon monoxide and nitrogen dioxide have now been brought down below international guidelines almost the

**Figure 2.30** Most cities face problems from traffic.

whole year round. Under new regulations, cars registered in Mexico City have to undergo a pollution check every six months and those that fail will be ordered off the road one day a week. When local government declares pollution emergencies, high polluting cars must stay off the road, and petrol stations and factories can be closed down for the duration of the alert.

Another approach involved planting eight million trees within the 4,000km$^2$ of the city. The trees cover an area of 148,000 hectares, with a third being in built-up areas. Apart from taking up carbon dioxide, trees help remove other air pollutants.

Another traffic-based solution to air pollution is to guarantee mobility through efficient public transport systems. Priority may be given to public transport, with lanes for the exclusive use of buses and trams contributing to a better quality of service. Buses may get priority at signals if they can be identified on the approach, e.g. through transponders that interact with traffic lights.

Several measures restrain the use and impact of the private car. These include traffic calming, including speed restrictions, road narrowing and pedestrian zones. Traffic restrictions deter private car use within a city and by enabling higher flows along roads outside the centre, air pollutants are more effectively dispersed.

Another way to limit private traffic is through market-based instruments such as congestion charging and road pricing, which is increasingly used in Europe.

Commuting trips account for the highest percentage in travel demand, mainly at peak periods. So, should businesses generating this traffic demand take responsibility for reducing it as well as the traveller? Demand management measures (for example, 'teleworking' and flexible schedules) may optimize modes of travel and distribute the trips throughout less congested periods. However, flexible working patterns may militate against the provision of public transport systems: as people travel at different times, public transport services are no longer economically viable. Teleworking is becoming more and more common as the technology improves, and is now a real option for several types of work. Videoconferences may also help reduce the need for travel on business. However, it is too soon to predict future impacts of these technologies and whether they are preferable in life-cycle terms.

## 6.4 Urban climates

Let's take another view of Earth from space. The thermal image confirms well-known effects (such as differences in temperature between land and water), as well as changes to local weather and climate patterns. Urban areas remain warmer than surrounding areas, effectively trapping heat (Figure 2.31).

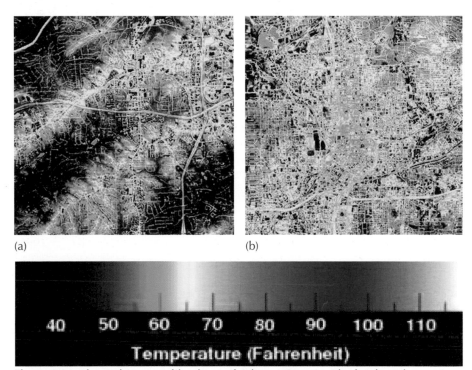

(a)                                              (b)

Figure 2.31 Thermal images, like these of Atlanta, Georgia. The bright red areas in these images are about 65 °C; dark green and blue areas are around 25 °C.
*Source*: www.nasa.gov/newsinfo/urban.html

Energy losses that we saw in Figure 2.9 and retention at night of the radiant energy that is absorbed by roads etc. during the day create 'heat islands'.

On hot summer days, urban air can be much warmer than the surrounding countryside. Scientists call this phenomenon the 'urban heat island effect' but you should not confuse it with global warming (**Blackmore and Barratt, 2003**).

There are a number of factors that contribute to the relative warmth of cities. During the day in rural areas, the solar energy absorbed near the ground causes water to evaporate from the vegetation and soil. So, while there is a net solar energy gain, this is compensated to some degree by evaporative cooling. In cities, where there is less vegetation, the buildings and roads absorb much of the solar energy input. Runoff is greater because pavements are largely nonporous, so there is effectively less water in the city and evaporative cooling is less.

Heat islands raise urban temperatures in the summertime, adversely affecting human health and the environment. Additionally, heat islands decrease urban ventilation, increasing air pollution levels and increasing the risk of heat-related illness and mortality. Heat islands also increase ground-level ozone pollution (**Blackmore and Barratt, 2003**). This is a harmful pollutant and the main constituent of summertime smog. Heat islands also increase energy use: higher temperatures resulting from the heat island effect can increase the demand for energy to cool homes, offices, and other buildings.

Thermal images, like those in Figure 2.31 can pinpoint energy losses and provide information that can improve tree-planting programmes in urban areas. Planting trees and bushes can help reduce urban temperatures as well as make cities greener. By providing shade and reducing urban temperatures, vegetation can save energy, but planting urban trees by design is essential. In the Northern Hemisphere, deciduous trees shade the south and west sides of buildings from the summer sun. Trees and bushes can also shade air conditioners, which work more efficiently when kept cool. Also, evergreen trees and bushes to the north can protect buildings from cold winter winds.

Trees grouped together create an oasis in a city and also cool nearby areas. Grouped trees also protect each other from the sun and wind, making them more likely to grow to maturity and live longer. Information such as this from satellites should help us make informed decisions necessary for sustainable urban growth and preserve our environmental futures.

## Summary

Urban development and the associated urban sprawl can lead to problems in the impact on natural habitats as well as in the growth of resource use. Efficiencies in resource use may be made by recovering waste for re-use. This avoids the energy and resource consumption for replacement materials. However, re-use can also apply to land, with brownfield development replacing the need to build on greenfield sites, thereby combating sprawl. Good urban design can also stimulate good environmental practice by facilitating industrial ecosystems in which wastes become resources for neighbouring activities. Good urban design also helps combat problems, such as the urban heat island, and has implications for energy use and pollution.

# 7   Conclusion

In this chapter we have looked at the role of technological responses to some environmental risks associated with urban development and their role in working towards sustainable development. However, these responses must be considered alongside social and economic considerations. Revisiting the key questions at the start of the chapter, an essential tool for managing environmental problems includes embracing whole product life cycles in an environmentally acceptable way. By doing so we can change our environmental futures by:

- designing goods and services that require less raw material and energy, perhaps by recovering resources from waste streams;
- fostering a robust economy for secondary (recovered) materials by creating policies that promote and support materials conservation, recovery and efficiency;
- using land efficiently to avoid urban sprawl with its associated risks, including the contribution of traffic to urban pollution.

Cities have advantages, but environmental considerations often ignore benefits such as economic efficiency and focus on problems. Traffic congestion, for example, is a problem to many but we may have different perspectives on it. Attempts to deal with the 'problem' will often fail if they do not address more difficult questions. Why do people need to travel? What needs do they seek to satisfy? What are the alternatives to more and faster roads? In addressing the questions posed in this chapter, we need to widen our boundaries beyond solving problems such as congestion and take in problem identification and definition. We must consider more than the technical aspects of urban life that have been the focus of this chapter, and include all stakeholders and power relationships. This highlights the difficulties inherent in such problems and the need for integrated approaches. We must embrace these wider issues if we are to change our environmental futures. In the following chapters such features are explored under headings of economic and political responses.

# References

Bingham, N. and Blackmore, R. (2003) 'What to do? How risk and uncertainty affect environmental responses' in Hinchliffe, S.J., Blowers, A.T. and Freeland, J.R. (eds).

Bingham, N., Blowers, A.T. and Belshaw, C.D. (eds) (2003) *Contested Environments*, Chichester, John Wiley & Sons/The Open University (Book 3 in this series).

Blackmore, R. and Barratt, R.S. (2003) 'The dynamic atmosphere' in Morris, R. M. et al. (eds).

Blowers, A.T. and Elliot, D.A. (2003) 'Power in the land: conflicts over energy and the environment' in Bingham, N. et al. (eds).

Brandon, M.A. and Smith, S.G. (2003) 'Water' in Morris, R.M. et al. (eds).

Burgess, J. (2003) 'Environmental values in environmental decision making' in Bingham, N. et al. (eds).

Castree, N.C. (2003) 'Uneven development, globalization and environmental change' in Morris, R.M. et al. (eds).

Drake and Freeland (2003) 'Population change and environmental change' in Morris, R.M. et al. (eds).

Furniss, P. (2003) 'Troubled waters' in Bingham, N. et al. (eds).

Hinchliffe, S.J. and Belshaw, C.R. (2003) 'Who cares? Values, power and action in environmental contests' in Hinchliffe, S.J. et al. (eds).

Hinchliffe, S.J., Blowers, A.T. and Freeland, J.R. (eds) (2003) *Understanding Environmental Issues*, Chichester, John Wiley & Sons/The Open University (Book 1 in this series).

Molina, M.J. and Rowland, F.S. (1974) 'Stratospheric sink for chlorofluoromethanes-chlorine atomic catalyzed destruction of ozone', *Nature*, vol.249, pp.810–2.

Morris, R.M., Freeland, J.R., Hinchliffe, S.J. and Smith, S.G. (eds) (2003) *Changing Environments*, Chichester, John Wiley & Sons/The Open University (Book 2 in this series).

Reddish, A. (2003) 'Dynamic Earth: human impacts' in Morris, R.M. et al. (eds).

Society of Environmental Toxicology and Chemistry (SETAC) (1993) *Guidelines for Life Cycle Assessment: A Code of Practice*, (first edn), Brussels.

# Answers to activities

### Activity 2.1

If the total population in developing countries grew from 1.7 billion in 1950 (a billion is $10^9$ or one thousand million) to approximately 4 billion in 1990, this represents a change of (4/1.7) = 2.4 times. If the urban population increased from 286 million to 1.514 billion so the change was (1.514/0.286) = 5.3. So urban populations are growing more rapidly in these parts of the world.

## Activity 2.4

My list would include the heating or air conditioning system in a home, the telephone system, the computer system at home or in a workplace, and a production system, for example the one involved in producing this book. We use the word 'system' in many ways. It may mean hardware, as in the heating or telephone systems, but equally it may be concerned with information or even people interacting with technology. 'System' also denotes any set of components identified as working together to achieve a purpose. Water sources, treatment and distribution are part of the water supply system of an urban area; there is also a parallel wastewater treatment system. We can view both as sub-systems of the water resource management system of an urban area. Similarly we have transport systems in and between urban areas and we also have 'ecosystems'.

## Activity 2.5

Your home provides for the focus for your basic needs of food, comfort and shelter. Depending on how you have broken down your list, it is likely to include specific food items (vegetables, meat, drinks, etc), all of which relate to the food production systems. Then there are inputs such as water supply, electricity, and the fossil fuels (gas, coal, oil) as well as all of your other purchases such as furniture, electrical equipment, clothes, building and decorating materials, and so on. All of these link to manufacturing systems as well as to the use of energy and materials in production. In addition there are your transport requirements for travelling around on business or for leisure. Then there are the outputs from your home, the sewage to the wastewater system, the food waste and packaging that goes to the municipal waste management system, not to mention the waste heat from losses through the building fabric.

## Activity 2.6

We can identify the following data:

| | |
|---|---|
| Industrial steam boiler | efficiency 90 per cent ($\eta_1$) |
| steam turbine | efficiency 45 per cent ($\eta_2$) |
| large electric generator | efficiency 90 per cent ($\eta_3$) |

This information means that, for the boiler, 90 per cent of the input energy appears as output. When connected to the turbine, 45 per cent of this 90 per cent is passed on to the generator, which in turn provides as output 90 per cent of this input. Overall efficiency ($\eta$) can therefore be calculated as follows:

$$\eta = \eta_1 \times \eta_2 \times \eta_3$$
$$= 0.9 \times 0.45 \times 0.9$$
$$= 0.36 \text{ or } 36\%$$

## Activity 2.7

For a household that altogether takes 16 minutes worth of showers each day, a low-flow showerhead can save (14.3 – 11) 3.3 litres/min × 16 min × 365 days per year = 19272 litres of water each year, plus the energy required to heat the water and the disposal implications of the waste water. Thus there are resource savings as well as lower annual carbon dioxide emissions from the energy use.

## Activity 2.8

Low-flow toilets save water, but they also save energy in several ways. Indirectly, electricity is saved by having to pump less water to buildings, and process and dispose less sewage water. Directly, flushing less water reduces the loss of heat and air conditioning energy, because a toilet's standing water typically absorbs this energy and then is flushed away. Toilets can use an average of 20 litres per flush, more than three times the amount required by high-efficiency toilets. Money follows all that water down the drain. But, make the flush too low and multiple flushes are needed, eliminating the benefits.

## Activity 2.9

For the motor car, you may consider the vehicle itself as a closed system with a system boundary drawn around it until you need to refuel it, whereupon you call upon the resources of the larger petroleum industry system. Your car system also has sub-systems, including the engine, fuel system and electrical system. However, you also need the road system on which to drive the car, and this is another part of the essential urban infrastructure. You must also operate the car (human system) within the legal system, and so it goes on.

## Activity 2.15

Our summary is as follows.

| Benefits | Difficulties |
|---|---|
| an integrated view – from 'cradle to grave' at it's best it can be an objective process indicates overall environmental impact | problem in deciding on basis for comparison – the functional unit defining the system boundaries difficulty in collecting some data for the inventory comparing different impacts – may involve subjective assessments |

# Chapter Three

# Economic analysis and environmental responses

Paul Anand

## Contents

# 1    Introduction: uncertain decisions

**Activity 3.1**

Imagine that you are the minister for environment with special responsibility for food issues. It is late on Saturday afternoon and the phone rings. The Department of Health has just heard from a doctor in a remote rural community who has just seen the fourth case of life-threatening food poisoning within a week. This might be a coincidence but all the victims worked in trades related to the beef industry. Then your mobile phone rings. This time it is your press secretary explaining that journalists would like a comment in time for the Sunday papers.

What do you say? Write down some lines of argument that you might take before reading the Introduction to this chapter.

Over the years, a number of ministers for agriculture and health both in the UK and around the world have been in very similar situations to the one described above. Perhaps most famously of all, in 1990 John Gummer (the then UK minister for agriculture, fisheries and food) handled the dwindling confidence in the safety of UK beef by feeding his daughter a burger to the delight of TV cameras and a barrage of criticism from the papers the following day (see Figure 3.1). At about the same time, Edwina Curry, at the time a junior minister for health, faced a similar problem when it was discovered that salmonella was becoming endemic in the UK chicken population. She was first to articulate the position on a prime-time TV chatshow in a way that was transparent and correct, and yet, within a matter of days, she was forced to resign. From these two cases, you might take the view that saying as little as possible, whether you think there is a risk to the human population or not, is the best course of action. But, of course, 'no comment' is not an option for a minister of state.

So what might you say and what should you do? Decision analysis, at its simplest, suggests that we begin by describing the options available and their possible outcomes, and that we pick the action that has the best prospects. Suppose you have three options, which range from banning the sale of UK-produced beef at one end of the spectrum (BAN), to doing nothing at the other (STATUS QUO). There might be a number of intermediate options you could consider but for the sake of simplicity we assume that there is just one, which involves introducing tighter quality control regulations on the production of food (REGULATE).

Next we develop an understanding of the variable and important states of the world that will affect the consequences of different courses of action. Here, I am going to assume that there are three possible states of the world that matter. First, these reported cases might just be evidence of a freak statistical event and not imply anything untoward about food safety, a state of the world I call 'Safe'. Second, they may indicate the existence of a new food risk issue that might affect

**Figure 3.1** John Gummer and his daughter eating burgers.

a small proportion of the population, a possible state I shall refer to as 'Low Risk'. Third, they may indicate an issue that might affect a large proportion of the population, a possible state I shall refer to as 'High Risk'.

Finally, we specify the values of the different outcomes that might arise for each action, given the different possible states of the world. To make things concrete, we'll measure these outcomes, for the moment, in terms of cases of serious illness or death, figures for which appear in Figure 3.2, which summarizes the decision problem.

Now that we have a description of the problem, we apply a rule that uses the information contained in the table to identify a best action. One rule often proposed, maximin, requires that for each action, we find the worst possible outcome (the minimum outcome); then we select the action that has the least bad outcome. In this case, the worst outcomes are 100 cases (BAN), 500 cases (REGULATE), and 600 cases (STATUS QUO); so the maximin outcome is 100 cases, which means that the best option is BAN.

But, if this were the situation, would you really announce an immediate ban on the consumption of UK beef? This might seem like a 'precautionary' approach but perhaps it is too precautionary? Because it gives weight only to the worst possible outcome, it might be thought to be excessively risk-averse (Anand, 1993). There are other approaches, and they push the decision in quite different directions. You might remember that John Gummer famously claimed that UK beef was

'No evidence of harm' and 'evidence of no harm' are discussed in **Bingham and Blackmore, 2003**.

| | cases of serious illness/death according to each action | | |
|---|---|---|---|
| | BAN | REGULATE | STATUS QUO |
| high risk | 100 | 500 | 600 |
| low risk | 100 | 200 | 500 |
| safe | 0 | 0 | 0 |

*states of the world*

Figure 3.2 The food-scare decision problem.

'safe', by which he meant that there was no 'scientific evidence' to suggest it was risky. He was using the term 'scientific evidence' in a particular but understandable and, for the time, accepted, sense for it was true that no causal chain from the eating of contaminated beef to the contraction of human BSE had been observed.

One might also object to the application of maximin because other relevant information has not been accounted for. Banning sales of UK beef for a long period could threaten the long-term future of the entire beef industry and challenge related industries. We might be concerned about the consequences for employment and rural economies around the country and ask whether the policy response being considered is commensurate with the risk to human health and life. These consequences could be reflected in the table too.

This activity indicates the role that economic frameworks can play in identifying relevant contributions of science and politics (further discussed in the adjacent chapters). On the one hand, natural science seems essential for understanding the actions, possible states and likely consequences available to us. However, option selection depends not just on a solid understanding of the relevant natural science but also on how we feel about risk, how likely we think some outcomes are, and the relative weights we want to give to different aspects of those outcomes. These things depend on our preferences and indicate the need for a process that indicates what peoples' relevant preferences actually are, be it political or otherwise.

In this chapter on the economics of policy responses to environmental issues, I shall be considering both the tools and concepts with which economists work (economic analysis) as well as the transactions and institutions that serve to allocate resources (the economy). You will see that the economic analysis of environmental policy is very much concerned with incentives and objectives as well as decision making. Given the aims and objectives that economic actors have, will policies provide incentives for them to do what we would wish? This is the question we shall be asking. Towards the end of the chapter, we shall also ask what our aims and objectives really should be.

The chapter is structured as follows. We have already looked at one framework for handling policy responses to uncertainty (maximin) and in Section 2 we examine a modelling approach that will help us understand decisions where human action affects environments. We shall consider this in the context of policies that aim to make competitive fishing sustainable, though it could be applied to the exploitation of any non-cultivated life forms.

In Section 3, we turn our attention to pollution and the debate between standards (command-and-control approaches) and economic instruments (market-based approaches). Although natural scientific research has often given rise to standards that specify particular pollution levels, we shall see that economic instruments bring additional advantages in terms of policy efficiency. In Section 4, we extend our analysis by looking at issues that arise in the context of international environmental problems. Many environmental issues involve more than one country but as countries are all powerful, in the sense of being sovereign, international agreements have to be self-enforcing. In the section, we shall use game theory to shed light on how self-enforcement might be brought about, or not, as the case may be.

Finally, in Section 5, I want to move to questions of well-being and sustainability. We shall note that some people argue a more pessimistic case for sustainable consumption than others. We shall also see that there are philosophical, psychological and even basic accounting reasons for thinking that conventional economic measures of consumption are not as closely related to well-being as their central role in policy implies.

By the end of the chapter, I hope you will understand that the economics of environmental policy demands that we are clear about objectives, that incentives must be right if policies are to succeed and that uncertainty can be a consideration in determining which policy is best.

# 2   Renewable resources: fisheries, exploitation and sustainability

Throughout the European Union (EU) fish stocks are in crisis. Once thought of as a never ending supply, fish are becoming scarce. In fact, almost two-thirds of commercial fish stocks are over-exploited. Some are even on the verge of commercial extinction ... The recent collapse of the cod stock in the North Sea has led to short-term closure of the fishery and has illustrated the need for recovery plans and long-term management measures. Cod and chips, pickled herring and bacalhao (salt cod) may not be familiar national dishes for future generations.

(WWF (World Wide Fund for Nature), 2001, p.2)

Understanding the exploitation of resources that can regenerate (renewable resources) plays an important part in the analysis of sustainability. Some renewable resources, such as solar energy, are not affected by our use or consumption of them. In contrast, economic activity that depends on our exploitation of wild, living organisms raises interesting problems, (see Figure 3.3). Because animals and plants have a capacity to reproduce themselves, and because this capacity is affected by the extent and manner of our exploitation, our economic models and the policy prescriptions that follow need to incorporate this interaction.

**Figure 3.3** Traditional fishing in Sri Lanka: it is often difficult to predict how sustainable, or otherwise is the exploitation of resources.

## 2.1 The bionomic fisheries model

Fishing resources are discussed in **Blowers and Smith (2003)**; **Smith and Brandon (2003)**; and **Castree (2003)**.

Models in this tradition are known as 'bionomic' because they seek to integrate insights from biological science and economics. Inevitably, they find it difficult to do justice to the detail of either discipline but, judiciously applied, they can be used to shed light on some general principles that apply to decision making and policy formulation with regard to renewable resources. Basically, the problem of fishing in open access waters is an example of the 'tragedy of the commons' (see **Castree, 2003**).

Figure 3.4 summarizes the basic elements of the neo-classical model of fisheries exploitation. Let me point out the main features. First of all, note the axes: the horizontal axis represents fishing effort, which reflects the numbers of boats, workers and other inputs involved; while the vertical axis is measured in monetary units. Now note that total costs (TC) increase proportionately with effort while revenue can actually decrease once a certain level of effort has been exceeded.

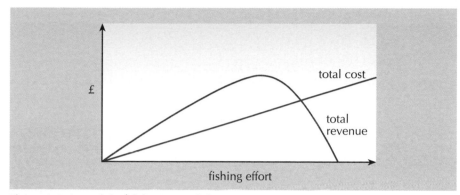

**Figure 3.4** A model of fisheries exploitation.

Because economic agents are assumed to maximize the difference between revenue and cost (i.e. their profit or 'rent'), we are interested in the difference between revenue and cost.

Total revenue can be defined as the quantity of fish sold multiplied by its market price. Typically, we assume that market price is fixed (reasonable if the fishery is a small part of the world market) so the curve reflects the relationship between the effort and catch. Initially, more effort results in larger catches, but there comes a point when the catch is exactly balanced by the ability of the fish stock to reproduce itself. As fishing effort increases beyond this point, fish are caught more quickly than the rate at which they can reproduce so the size of the stock and the catch enter a spiral of decline. The relationship between total revenue and effort is usually curved and tails off more quickly than it rises.

The relationship between TC (the cost of fishing) and effort is usually assumed to be a straight line. Again this is an approximation that abstracts from some important facts (like technological progress) but it reflects the fact that average costs don't rise or fall much as capacity is increased in the short term. In this respect, the cost structure of fishing is quite different to manufacturing industry where average costs often fall dramatically as effort (output) increases.

Using this analysis, we can identify four significant levels of effort (see Figure 3.5). A is the level of effort that maximizes economic profit ('rent' is strictly the correct term). B is a higher level of effort that maximizes revenue from the fishery, which may be close to the maximum sustainable yield (the most that can be physically caught on a regular basis without undermining the reproductive capability of the population). Fishing effort beyond this point results in a fall in the sustainable catch. C is the level of effort that should result from a competitive market with open access – new capacity will be added to the fleet so long as a profit can be made. In theory, according to our simple graphical model, open access to

Drake and Freeland (2003) discuss maximum sustainable yield

fisheries should not see levels of effort higher than C, but in reality we do see fish stocks being fished out suggesting that effort levels reach point D.

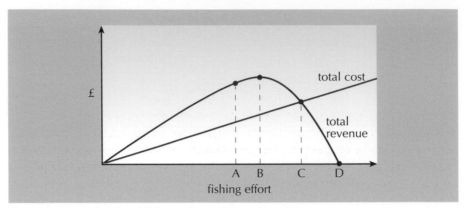

**Figure 3.5** Four levels of fishing effort.

So why is this? There may be a number of reasons but an important one is uncertainty. Boat owners often have very little idea of the exact shape of the total revenue curve, which shows the relationship between fishing effort and total revenue. If it is steep, then the difference between point C, the open access level of effort and point D, the point at which the fishery collapses, might be small. In that case, it would be easy for competition to pick level of effort D by mistake.

To combat these problems, policy makers have devised a number of standards (on inputs or catches) and economic instruments with which they can address the problems of over-exploitation. In the next section we shall see how well these policy options work.

○      What level of effort would prevail in a competitive open access fishery?

●      In theory, fishing effort continues until total revenues equal TC (level C on Figure 3.5). However, in practice, overfishing may result in effort levels above this point, particularly if there is uncertainty about the point at which additional effort results in zero or negative profit.

## 2.2  Standards and economic instruments for sustainable fisheries

To reduce the risk of extinction of the resource, fisheries regulators have devised a range of policy instruments for restraining fishing effort (see Table 3.1).

**Table 3.1**   Standards and economic instruments

| Standards (including limits) | Economic instruments |
| --- | --- |
| closed seasons | taxes on catches |
| gear restrictions | individual transferable quotas |
| limiting vessel numbers | |
| aggregate catch quotas | |

Not all of these approaches, however, are equally desirable in terms of economic efficiency. The closing of fisheries during breeding seasons may be essential for biological purposes but has no impact once grounds are open again. Alternatively, restrictions can be imposed on the size of engine or fishing gear used. This may reduce total catch but it does so by forcing the use of equipment that is smaller in capacity than would otherwise be the case. This in turn forces costs up, and profit to the industry down. It is important, however, that we achieve environmental goals at least cost. Failure to do so means there may be other policies that would produce the same environmental outcomes with less impact on human welfare or the environment.

Another approach to regulation limits the number of vessels allowed to access fishing grounds. Unfortunately, this too is inefficient as it leads to a phenomenon known as 'capital stuffing' (Tahvonen and Kuuluvainen, 2000), whereby the capacity of each boat is increased through a variety of means including use of better fish location equipment, faster and bigger net handling and larger crews.

Some fisheries regulators, like the Finnish authorities responsible for the Baltic Sea salmon fishery, have tried to impose an overall quota on the catch but again the policy does not encourage efficiency. As soon as the fishing season opens in the summer, boat owners have an incentive to maximize their share of the overall catch. Consequently, the incentive is to buy faster boats than would otherwise be necessary. In the Baltic, there is the additional problem that the aggregate quota is fished out before Christmas at which time salmon have to be imported from Norway, just when fish prices happen to be highest.

Given these constraints, many economic models suggest that it is only taxes on catches or individual transferable quotas that are usually efficient. If the tax is correctly calculated, it reduces the revenue received from fishing and thereby produces the desired level of fishing effort. The political drawback is that tax might be transferred out of the industry and/or area so is likely to be resisted by the industry. **Individual transferable quotas** (limits on the amount one person can catch but that can be assigned to others for a price) can also achieve the desired level of fishing effort without encouraging inefficient production activity. They act like property rights over the fish and because of transferability, they are bought up by those who are most efficient at fishing; we come back to this type of arrangement in Section 3.

individual transferable quota

In many situations, economics prefers economic instruments over standards, a point we shall come back to and develop in Section 3. This is not, I should stress, an ideological result but simply the result of an analysis of how best to achieve a particular level of production.

## 2.3 Fisheries policy in practice

At the time of writing, the 20-year-old European Common Fisheries Policy (CFP) was being completely reformed with the prospect of many thousands of jobs in fishing being lost. The CFP had been formulated on insights consistent with the bionomic model discussed above, so what, one might ask, went wrong? A few substantial problems illustrate difficulties that exist both in Europe and around the world:

- Although aggregate annual quotas for key species are set by the CFP, their level, after political negotiation between countries, is often higher than that advised on scientific grounds. Politicians appear to be influenced by industry pressures to maximize short-term catch levels – even though it might seem counter-intuitive to not want to protect long-term gains.
- Many fishing industries benefit from government subsidies. In the EU, there have been grants for taking boats out of service but there are also socio-economic funds that support the introduction of new boats. There is no getting round the fact that this is a crazy state of affairs (see Figure 3.6).

**Figure 3.6** In the EU grants exist simultaneously for introducing new fishing boats and decommissioning others.

- A lot of deep-sea fisheries are still completely unregulated. As stock depletion and competition pushes boats to find new grounds, they come across new species whose biologies are particularly susceptible to overfishing. The orange roughy (*Hoplosthetus atlanticus*), for example, is a late developer that does not start breeding until its mid 30s so any sudden collapse in stock numbers will take decades to replenish.
- There are international agreements, such as the UN convention on biodiversity and the UN Fish Stocks Agreement, which impose on EU fleets a duty to fish sustainably in non-EU waters. However some EU agreements with Western African and Latin American countries both fail this requirement and result in competition with local small-scale fishing fleets.

In short, the policy problems of how to address over-exploitation of common resources depend both on narrowly defined economic and biological factors such as those identified by the traditional bionomic approach and broader political-economy issues to do with factors such as the provision of government subsidies and international equity.

## Summary

- A bionomic analysis of fisheries' sustainability and exploitation combines biological and economic insights.
- Using this analysis, it is possible to specify different levels of fishing effort and their economic and biological consequences.
- Both standards and economic instruments are used to regulate fisheries in practice.
- Threats to sustainable and fair exploitation of fisheries arise from industry subsidies and failure to set adequate controls on catch sizes.

# 3   The regulation of externalities: standards and economic instruments

## 3.1 Pollution as market failure

While fishing provides an example of a situation where a producer's action can have negative effects on everyone, the producer included, pollution provides an example where the person producing the pollutant may, or may not, be concerned about the negative consequences of their actions. Acid rain provides a good example (see Box 3.1, overleaf).

The costs of such pollution go unpaid by the burners of fossil fuels or users of agricultural fertilisers largely responsible for them. Costs (or benefits) such as these, when caused by economic activity but not reflected by the pricing system, are called **externalities**. They are a ubiquitous feature of economic activity and represent a fundamental problem for free-market processes because, in their

externalities

presence, markets will not lead to the right quantities of consumption or input use.

## Box 3.1 Acid rain

In the northern hemisphere, acid rain, caused by emissions of sulphur dioxide, oxides of nitrogen and ammonia, is a major source of environmental degradation. Aquatic environments of Atlantic salmon and brown trout in the UK have been threatened as has the biodiversity of mountain areas. Forests suffer as nutritional deficiencies in trees are exacerbated and some species, like the red spruce, are believed to suffer direct damage to leaves or needles. In addition, acid rain impacts on human health, agricultural crops, freshwater eco-systems and building materials and one estimate indicates that the damage to Europe was valued at over $23billion (Alcamo et al., 1990).

## Activity 3.2

What policies, if any, would you introduce to deal with the negative externality suffered by a salmon farm arising from pollution by a factory upstream? List some possibilities before reading on. For an answer, go to the end of this chapter.

## 3.2  Choosing policy instruments

There are many situations in which economic activities generate externalities that could be either negative (costs) or positive (benefits – see Chapter 6) and in this section we look at the economic analysis of remedial policies that aim to reduce the effects of negative externalities. The best response depends, as always, on the objectives and the details of the situation in hand. It is customary to distinguish, as we have already noted, between standards (specific legal limits specified in quantitative terms) and economic instruments (such as taxes or tradable pollution/exploitation permits). We have already seen standards and economic instruments in the context of fisheries and renewable resources and, as before, we shall see that economic analysis tends to favour policies that have strong incentives for compliance and relate abatement to its benefits.

## Pros and cons of standards

Sometimes referred to as 'command-and-control', this approach depends on the issuing of standards concerning amounts of pollution or types of production techniques that can be used. This approach is reflected by policies such as the UK's Best Practicable Means (BPM) or the USA's Best Available Control Technology (BACT), which require firms to use appropriate production methods as ways of limiting pollution.

On the positive side, regulatory approaches such as these fit in closely with the way in which policy in related areas, such as health, safety and work, is implemented. Governments are used to establishing laws and government bureaucracies are organized around the implementation of these standards. Furthermore, where there is a need to prevent changes that might be irreversible – loss of biodiversity for instance – then prohibitive standards may be necessary. Finally, standards are a good way of achieving a target outcome, so long as those standards are adequately enforced.

On the negative side, the enforcement of environmental standards often isn't very strong as the lack of enforcement officers makes the probability of being caught often quite low. Moreover, when offenders are caught, fines for non-compliance can be too low to act as an effective deterrent. In general, the regulators and regulatees negotiate the setting of regulatory standards and, inevitably, this can lead to them not being as stringent as they might or should be. This negotiation exemplifies the concept of 'regulatory capture' that was developed by North American economists. It applies particularly to bureaucracies that are open to direct political influence as they are in the USA. Where bureaucrats are more protected from political influence, as is the case in Europe, there is less reason to believe that regulators will be captured by the industries they are supposed to oversee.

Over time, technology may develop more quickly than standards unless the standards are specifically designed to anticipate change (as was the case with BACT and BPM legislation). The state of play at the time of writing is that many environmental standards worldwide are being overhauled to overcome the problems mentioned above. Some countries are pursuing new policy instruments such as taxes or voluntary agreements. In other cases, regulatory reform is being based on life cycle analysis, extending producer responsibility and a greater involvement of stakeholders.

These shifts are sometimes labelled 'ecological modernization'. See Chapter Six for more on this political shift.

○   List two benefits of the 'standards' approach to regulation.

●   Regulatory standards a) fit in well with the way in which governments establish laws and b) are well suited to the achievement of environmental targets.

## Economic instruments I – green taxes as a method of reducing externalities

One method of making a firm take account of (or 'internalize', to use the jargon) negative environmental externalities is to levy a tax that requires that the polluter pays. So long as pollution emissions are easy to measure, such taxes can easily be calculated. Because a tax raises the cost of pollution, polluters have an incentive to abate (reduce) the amount they pollute.

If one knows what the 'right' amount of pollution is, from society's perspective, and what the costs of additional abatement are, then it is possible to set a tax so that the 'right' level of pollution is produced. (Note that it is rarely worthwhile eliminating all pollution and, indeed, zero pollution is technically and theoretically impossible.)

Taxes appropriately set are capable of achieving the same outcome as standards, so taxes have no advantage in terms of pollution outcomes, per se. Nonetheless, taxes do have some advantages over and above standards. First, they help to minimize abatement costs. If abatement costs are different across firms, the uniform imposition of a single standard will not, in general, minimize abatement costs. With pollution taxes, however, each firm reduces its pollution up to the point that makes sense in terms of its costs and tax levels. Firms who find it costly to reduce pollution do little abatement and pay a large amount of tax. Firms with low abatement costs do much more abatement and pay less tax. And, because those for whom abatement costs are highest do least of it, the total costs of abatement are minimized. (This may not be the only consideration if we are concerned about the distribution of abatement activity, which for symbolic, political and ethical reasons we might be. For example, this may create 'dirty zones' where high pollution is tolerated – see **Blowers and Elliot, 2003** for a discussion of uneven exposure to pollution.)

Once a firm has reached a particular emission standard, it has no further reason to abate but a tax provides two kinds of incentives to do more abatement. Firstly, if due to technological progress, abatement becomes cheaper, the level of abatement will rise. If a standard had been used instead, technological progress would have just made it cheaper for the firm to achieve the standard. Second, and related, a tax provides a stronger incentive than standards to develop technologies that reduce abatement because new technologies result in reduced abatement costs plus savings in tax paid.

Finally, it is also worth pointing out that such taxes can be earmarked for environmental improvement, a practice often referred to as 'recycling'.

○    What do we mean when we say that environmental taxes are aimed at getting firms to internalize their externalities?

●    Environmental taxes are designed to make firms take account of the costs of pollution in their decision making, even though the market would not make them do so otherwise.

## Economic instruments II – pros and cons of tradable permits

Another approach to dealing with pollution problems created by externalities involves the use of tradable permits. The idea, proposed by a Canadian economist, John Dales (1968), is just that permits allowing a specific amount of emissions are issued to polluters who are then free to trade the permits. Because the government issues these permits, it can control, in principle, the overall amount of pollution. Furthermore, tradeability means that permits will go to those who can extract most economic value out of them. Other things being equal, this means that permits will go to producers who create relatively high economic value for relatively little pollution cost, so tradeability plays an important role in reducing pollution relative to economic activity. For this reason, tradability of permits ensures that abatement costs will be minimized. Another advantage is that firms are not prevented from entry into new markets, as they might be under a fixed licensing system. If a business develops a new product that results in pollution, then that business can bid for permits on the open market. So tradable permits are particularly appropriate for relatively dynamic economies.

However, permits suffer from a number of practical problems. There is, for one thing, growing concern about the distributional consequences of the initial allocation of permits – the so-called 'grandfathering' problem. One way of distributing permits is simply to give them to companies who, traditionally, have been engaged in the activity concerned. When these companies come to sell their permits, they could enjoy an unexpected source of income though such 'windfall' gains, or profits can be taxed as they were when the USA introduced tradable permits to cover the ozone-depleting substances.

In practice, complexity, administrative costs and political resistance also raise problems for this approach. Where the environmental issue has been understood for some time there tend to be existing standards that complicate the design and operation of an additional permit system. By contrast, newer environmental concerns can be subject to fewer pre-existing standards making the introduction of permit trading somewhat easier (see Figure 3.7). There may also be political resistance to the introduction of such systems on the grounds that it is unethical to trade in pollution, notwithstanding the environmental advantages of introducing economic signals and incentives into the abatement process.

A number of permit trading systems now exist around the world, though they tend to be confined to air protection and water resource management in the USA. (A notable exception is the protection of fisheries, which makes use of tradable permits in Australia, Canada, Iceland, Netherlands, New Zealand and the USA.) Early attempts to establish trading schemes in the USA were not particularly successful, though developments in the 1990s appear to be having more impact. Of particular relevance is the market for sulphur dioxide emissions established by the 1990 Clean Air Act aimed at reducing acid rain problems. The program has a 'cap and trade' approach: each year, a cap on annual emissions is fixed so that emissions will, by 2010 be reduced to below 1980 levels. Of these allowed emissions, 2.8 percent are auctioned to new or current polluters. The price of these permits has fallen dramatically since trading first began in 1998 suggesting that firms have already found ways of reducing emissions.

Such schemes are likely to become more, rather than less, important, though the political problems associated with their implementation should not be under-

**Figure 3.7** Today's environmental concerns are subject to fewer pre-existing standards, so permit trading maybe somewhat easier.

estimated. An international trading scheme for $CO_2$ emissions is being developed following the Kyoto agreement (see Chapter Five) but, like all such schemes, it will require that pollution activity can be clearly identified and measured, and that permits are legally secure (i.e. won't be revoked by some new political party). The greater the risk that the permits scheme will be revoked, the less reason there is for trading to take place.

The design of policy instruments continues to evolve and many observers detect a move towards the adoption of economic instruments. The use of tax and subsidy provides policy makers with the ability to intervene in markets while pursuing policies that maintain a link between benefits and costs. Where issues of distribution are concerned (i.e. who pollutes and how much), or where outright prohibitions are deemed necessary, regulations may be more valuable. By way of summary, the pros and cons concerning economic instruments and standards discussed are summarized in Table 3.2.

**Table 3.2**   Pros and cons of environmental policies

|      | Economic Instruments | Standards |
|------|----------------------|-----------|
| **Pros** | Can minimize abatement costs. Encourages adoption of clean technologies. Provides revenue which can be used for environmental purposes. | Good for reaching targets (if effectively enforced). Fit current approach to government intervention and legislation. |
| **Cons** | Allows polluters to pay their way out of trouble. | Fixed standards may become outdated. Variable standards may be insensitive to cost. |

# 3.3 Case study: regulating the water industry

Public water supply involves taking ('abstracting') untreated water from underground and overground sources. The water is then treated and distributed via a network of pipes to consumers who make greatest use of it in washing and the flushing of lavatories. Broadly, there are three reasons for regulating the supply of water. First, there is a natural monopoly in the distribution of water (a natural monopoly is a term used to describe a situation where one provider or producer dominates a market for reasons that are a result of the technical structure of the market); usually we only need one network of pipes supplying water to households. Second, most consumers are not well placed to directly determine the quality of their water supply. One cannot determine the amount or type of bacteria in water that comes from a tap merely by looking at it, so the consumer suffers from not being able to judge the quality of the product being supplied. Third, significant externalities arise from the supply of water. Effluent from sewage plants devoted to the cleaning of waste water can pollute the watercourses into which such water is discharged (though the effects are complex and depend on the amounts of dissolved oxygen and bacteria, as well as the volume and temperature of the discharge).

## Activity 3.3

In 1998, UK industrial sources of effluent had to obtain 'consents' and 'authorizations' from the national Environment Agency (EA). Consents may specify time, place, volume, temperature and composition of discharge but are not sensitive to abatement costs nor are they tradable. Charges for these consents (what I have been calling standards in this chapter) have been levied since 1991; they are designed to cover the costs of administering the system and are based on the maximum discharges allowed. Given the UK's policy, what would you do to make it more efficient?

## Comment

1   Perhaps the most crucial point about the UK's policy at the time is that because consent charges were set up so as to cover administration charges, the costs of the externality (pollution) are not correctly reflected on the polluters of water. By contrast, France, Germany and the Netherlands all operate combined schemes in which there are direct standards and taxes. Introducing taxes in the UK would help set the right level of pollution by allowing polluters to take account of the cost of pollution as well their own costs of abatement.

2   In general, tradability of permits results in them being bought up by those who can make best use of them, so the UK should consider making the consents tradable. Regulators in the USA tend to prefer tradable permits to taxes. However, because the permits relate to a particular watercourse, there are often very few buyers and sellers and, in such 'thin' markets, the benefits of a competitive market mechanism may not emerge.

We should remember that efficient policies are better placed to reduce abatement costs or increase environmental quality than inefficient ones. Moreover, there is a potential interaction between pollution and water abstraction: one method of dealing with the externality caused by pollution is to merge companies responsible for sewage treatment and provision of water supply. In that case, pollution becomes an internal cost that is met by the merged firm. Finally, it should be noted that there has been some discussion of the desirability of combining tax and regulatory approaches but that the outcome of such a combination depends upon the precise manner in which the approaches are combined.

## 3.4  Promotion of clean technology

Finally, an idea that is attracting increasing attention involves the promotion of so-called clean technologies, i.e. those that do less environmental damage when used (see Chapter Two). Promotion might take the form of standards mandating use of newer technologies. Alternatively, it might involve subsidization and could range from giving consumers low interest loans (to purchase solar cells for example) through to support for research and development. In other words, the promotion of clean technologies might be seen as the constructive counterpart either to standards limiting use of 'older' or dirtier technologies or to the taxation of their pollutants.

Reasons supporting the development of this approach are linked, in part, to experience with previous attempts to regulate pollution. In a number of high-income countries, especially the USA, many of the easy pollution targets (i.e. those where pollution is emitted from a fixed point) are already subject to regulation. Furthermore, attempts to regulate pollution can, on occasion, just move the pollution somewhere else or give rise to conflicts between polluters and regulators. Clean technology promotion, by contrast, is more preventive than curative and the technology itself attempts to tackle the problem at its cause. Given the manufacturing potential, it is also welcomed by those parts of industry responsible for developing clean technologies.

## Summary

- Externalities are bads (negative) or goods (positive) whose production is not fully reflected in the price mechanism. Environmental examples range from air pollution to loss of scenic views.
- Historically, externalities have been addressed either by the imposition of standards or by the development of economic instruments like taxes or tradable permits.

- Standards may be good for achieving specified outcomes, but usually take insufficient account of costs and benefits of compliance and may be slow to adapt to changing circumstances.
- Market instruments, such as taxes or tradable permits, are usually preferable with respect to economic efficiency. However, their introduction requires information and skills that agencies might not possess and can attract a variety of political objections.

# 4   International environmental agreements

## 4.1   The Montreal Protocol as a success story

Though chloroflurocarbons (CFCs) have been used since the 1930s it wasn't until 1974 that it was shown that they might deplete the ozone layer in the upper atmosphere (a layer that filters out harmful ultraviolet rays from the Sun). In 1977, the USA started to limit their use as aerosol propellants, though it was not until 1986 that substantial effects were scientifically documented. Subsequently, the Montreal Protocol was signed in 1987 by 24 nations in which they agreed to a 50 per cent reduction of CFC production from the 1986 baseline. In addition, international trade in CFCs was banned. China and India, significant beneficiaries of low-cost refrigeration (which uses CFCs), were not signatories though a number of low-income countries were and they were given a 10-year period of grace before they were required to start reducing their CFC production. To help them do this, a $260m adjustment fund was established in 1990. Although a relatively small proportion of the world's nations (who were responsible for most of the CFC products in the world) was included in the agreement, it included the top six CFC producers.

In 1992, new scientific evidence indicated that an accelerated phase-out programme might save approximately one million lives a year from reduced skin cancers and the adjustment fund was then boosted to US$500m. At the same time, cleaner alternative technologies were being developed and the regulation of CFCs boosted market opportunities for companies involved in the development and sale of substitutes.

### Activity 3.4

The international reduction of CFC emissions following the Montreal Protocol was judged a success. Based on the description above, write down factors that you think might have contributed to this success. (For five suggested answers, go to the end of this chapter.)

## 4.2 A game-theoretic treatment of international environmental problems and remedies

Pollution is no respecter of national boundaries and many environmental problems have multinational or international aspects to them. Countries may sign international agreements to tackle problems caused by acid rain or global climate change, but how are these agreements to be enforced? As we shall see when we return to the economics of responses to global climate change, enforcement, and therefore success, is by no means guaranteed. In this section of the chapter, I want to highlight a set of tools – this time from game theory – that are increasingly used to understand how best to formulate environmental policies at international level.

### The prisoner's dilemma game

Game theorists have specialized to date in studying the interactions between rational players; to give you a flavour of this analysis and its insights, I'd like you to look at Figure 3.8.

|              |        | country Y       |                 |
|--------------|--------|-----------------|-----------------|
|              |        | abate           | pollute         |
| country X    | abate  | 2 , 2           | 0 , 3           |
|              | pollute| 3 , 0           | 1 , 1           |

Figure 3.8 Pollution abatement as a prisoner's dilemma game.

In Figure 3.8, we have two countries, X and Y, each of which has two strategies: they can either pollute or abate. Their payoffs from doing so, are measured here as a percentage of **GNP (Gross National Product).** These are summarized in the table.

GNP (gross national product). This is defined as the economic value of the returns to resources owned by a nation. It can be measured by calculating the value of total output, income or expenditure.

In each cell, the figures represent payoffs to country X and country Y respectively. So, for example, in the cell in which X pollutes and Y abates, the figures indicate that the payoff to X is 3 (per cent growth) while the payoff to Y is 0 (per cent growth).

This particular game is an example of the 'prisoner's dilemma' (PD) game, which refers to problems that arise when people can't make binding agreements. The essential element of PD games is that if each player (country in this case) acts in its own selfish interest, the outcome for both parties is worse than if they agreed to do something else. The PD game and its application to environmental responses has made a very significant contribution to economic thinking because, since Adam Smith's work in the eighteenth century, many people had assumed that economies in which individuals pursue self-interest would lead to desirable outcomes. The PD game demonstrates, technically, that the individual pursuit of self-interest does not necessarily lead to the best outcome.

See **Castree, 2003** for a discussion of this in relation to the tragedy of the commons.

## How should countries play the game?

Consider what happens when country X looks at its payoffs. It sees that it gets either 3 or 1 if it pollutes, depending on what action Y takes. In both cases, the payoff is better than X would experience if it abated, in which case the payoffs would be 2 or 0. So X should, rationally, pollute – this is called its dominant strategy. Similar reasoning applies to country Y (because the table is symmetric with respect to countries), so acting on their own, each chooses to pollute and ends up with a payoff of 1. But of course, if they could make an agreement to abate, and stick to it, then each would do better and get a payoff of 2.

The structure of the PD game (only the rankings of payoffs matter) is such that it characterizes, simply, an uncertainty that is common in environmental responses. In reality, there are often more actors involved but the nature of the dilemma they face remains the same.

### Activity 3.5

By rewriting the strategies and reinterpreting the numbers as gains from fishing effort, can you apply the PD game to the problem of over-fishing discussed earlier in this chapter? (Hint: assume two countries are exploiting a single fishing ground, and you may want to adjust the numbers slightly to reflect the different resource-use situation.) For an answer, go to the end of this chapter.

## And how might countries agree to co-operate?

The PD game suggests that getting agreement is important, but also that countries might have incentives to unilaterally deviate from any agreement; this makes implementation difficult. Furthermore, payoffs in reality may often not be symmetrical as in the PD game above. For example, low-income countries might suffer severe economic losses from the costs of using new, clean technologies developed in high-income countries that would gain from their adoption. To deal with problems such as these, it is possible to compensate those who might lose out from joining an international agreement. Game theorists call such compensations 'side-payments', an example of which would be that those who benefit from reductions in the production of acid rain compensate those who undertake those reductions.

As it happens, side-payments are not often used in international agreements and equity is often cited as a reason for this. One reason for this is that, where a large number of countries are party to the agreement, how compensation is to be distributed and funded fairly is not clear. However, we should also note that when the number of participants is relatively small, as in the Montreal Protocol, it has been possible for high-income countries to compensate low-income countries for the costs of participating.

Another approach involves linking issues. If countries combine the payoffs from two games (say for example one to do with environmental quality and a second relating to international trade) and the payoffs are right, the incentives for countries not to enter into and abide by an agreement are reduced. However, linkage is not problem-free. Firstly, it raises the costs of decision making and when payoffs are not identical for both parties, countries do not have the same incentives to co-operate. Furthermore, there may be other agreements that make linkage different – for example, the 'most favoured nation' clause of the World Trade Organisation means that any preferential trade relations offered to one country should be offered to all other WTO members.

See **Maples (2003)** for a discussion of trade agreements and environment.

A third approach, with notable precedents in the Old Testament, is retaliation ('eye for eye, tooth for tooth'). In international environmental agreements the point is not that we want to punish those countries that renege on any agreement but rather that we want to prevent them from doing so. For prevention to work threats of punishment need to be credible but there also needs to be some path back to agreement. Both points indicate the importance of being able to play the game more than once.

A fourth issue relates to national reputation. It concerns the observation that once nations commit themselves to agreements they tend to abide by them, reneging only rarely and then usually because compliance is impossible. Chayes and Chayes (1991, p.320) summarize this line of thought rather nicely:

> As in other relatively small communities of actors in continuing and thickly-textured relationships, members of a treaty regime have available a wide range of informal pressures and inducements to secure compliance with community norms. States have dealings and continuing relationships with each other over a range of issues. Questions of treaty compliance arise in an environment of diffuse reciprocity ..., with manifold opportunities for subtle expressions of displeasure, suspicion and reluctance to deal with treaty-violators in other contexts. A reputation for unreliability cannot be confined to the area of activity in which it is earned.

In short, there are a number of general principles that might be applied to the formation of self-enforcing agreements between nation states. Hopefully, these ideas give you a sense of the importance of incentives in the design of international agreements and a background against which we can return to one of the hardest challenges for environmental policy makers: global climate change.

○   The prisoner's dilemma game when applied to the analysis of pollution abatement suggests that nation states might have difficulties avoiding outcomes that are undesirable for both countries. Why might this dilemma not emerge?

●   The offer of additional (side-) payments; the linking of environmental issues to other agreements; the threat of retaliation; and the need to develop and protect reputation.

## The economics of responses to global climate change

At the time of writing, the main agreement concerning global climate change (and rising sea levels) was that concluded in Kyoto in 1997. As Barrett (1998) noted soon afterwards, the economic analysis of self-enforcing agreements raises grave doubts about the protocol's ratification and implementation. Whether Kyoto is implemented or not, the details of the agreement provide a useful vehicle for examining issues that must be addressed if countries are to have strong incentives to ratify and implement substantial reductions in greenhouse gas emissions (see Figure 3.9).

Figure 3.9 Countries need strong incentives if they are to begin investing in environmental conservation.

In what follows, therefore, I want to examine issues surrounding the global climate change response as revealed by analyses of the Kyoto protocol, particularly those points that deal with efficient abatement and effective implementation.

# Will emissions reduction be cost effective?

In order for $CO_2$ reduction to be economically worthwhile at a global level, benefits must outweight costs. However, as the agreement was framed, the planned distribution of abatement was not least cost. For instance, to shift a tonne of emissions abatement from an OECD (Organization for Economic Co-operation and Development) country to a low-income country could save US$100. (One estimate suggested that it might be possible to reduce by 90 percent the global costs of the agreement.) In part, whether these savings will be realized depends on whether the country quotas become tradeable over time or not. Tradeability helps to ensure that abatement is implemented by those who can do it most efficiently but European politicians have resisted tradeability on the grounds that it seems to allow rich countries to buy their way out of duties to protect the environment. And even if Kyoto were effective, the increased costs of producing goods and services faced by signatory countries (see Table 3.3) would mean that consumers will tend to move their purchases to non-signatory countries (a problem sometimes referred to as 'leakage').

**Table 3.3:**   Kyoto Annex I countries

| Signatory country | Kyoto reduction |
|---|---|
| Iceland | 110 |
| Australia | 108 |
| Norway | 101 |
| New Zealand | 100 |
| Canada, Japan | 94 |
| United States, European Union | 93 |
| Austria, Belgium, Denmark, Finland, France, Germany, Greece, Ireland, Italy, Liechtenstein, Luxembourg, Monaco, Netherlands, Portugal, Spain, Sweden, Switzerland, UK | 92 |
| **Economies in transition from centrally-planned to market economies** | |
| Bulgaria | 107 |
| Czech Republic | 92 |
| Estonia | 92 |
| Hungary | 110 |
| Latvia | 92 |
| Lithuania | 92 |
| Poland | 108 |
| Romania | 107 |
| Russian Federation | 100 |
| Ukraine | 100 |
| Slovakia | 92 |

Note: The table shows, using index numbers, the extent to which the Kyoto agreement allows or requires change in $CO_2$ emissions for signatory countries compared with their emission levels in 1990. For example, the agreement allows Iceland to increase its emissions by 10 per cent but requires Canada and Japan to reduce their emissions by six per cent.

## Will countries join the agreement?

The question of compliance, which in the case of global climate change means agreement, and then enforcement, is crucial. International agreements pose an interesting challenge for mainstream economic theorizing, which focuses heavily on the importance of incentives and the scope for free-riding (see Figure 3.10). Typically, international agreements have few if any sanctions for non-compliance so the incentives for free-riding are, on the face of it, significant.

**Figure 3.10** Do international agreements allow for free-riders?

However, in practice, non-compliance by signatory nation states is believed to be extremely rare (as the quote from Chayes and Chayes suggests). One might, therefore, conclude that although Kyoto does not stipulate sanctions for non-compliance, this is likely not to matter in practice as countries will comply because that is how sovereign states behave. Furthermore, if one compares Kyoto with the Montreal protocol that preceded it, it appears that Montreal succeeded without resort to the use of sanctions. However, because Montreal banned trade of CFCs between signatories and non-signatories, it provided strong incentives for countries to sign up to the agreement: no such trade sanctions exist in Kyoto.

One mechanism for encouraging participation and enforcement uses a minimum threshold to kick-start the agreement. For the Kyoto agreement to come into force, it has to be ratified by at least 55 countries also responsible for production of 55 per cent of the total carbon dioxide emissions of all the Annex I countries.

Close to the threshold, countries have good reason to ratify – their own behaviour might well trigger substantial action that will deliver significant good in terms of reduced global climate change, albeit a public good from which non-signatories also benefit. However, suppose that both the 55 countries and the 55 percent thresholds are met. At this point, the incentive for further ratification is rather weak – a substantial number of countries are already committed to taking substantial action. A country that ratifies Kyoto after it becomes binding has a limited impact on global abatement but bears its share of costs. Such countries are, in a sense, acting unilaterally.

## Will countries invest in cleaner technologies?

Another potential problem involves the investment in abatement or clean technologies. It may be that countries find themselves in a game of 'chicken', where every nation waits for every other nation to make the first investments. Of course, if all the signatories wait for others to move first, then no investment will take place. Kyoto dealt with the issue by requiring each Annex I country to demonstrate progress in achieving its target by 2005. However, if a large number of countries make little progress, renegotiation of an agreement's timetable seems more credible than punishment. Indeed, who would be left to implement the punishments?

Barrett suggests that what Kyoto lacks is an incentive mechanism to ensure full participation. Montreal provides strong incentives to participate and comply by banning trade of CFCs with countries that fail to participate but, as Barrett points out, virtually all goods contribute to global climate change so banning trade in them is not feasible. Chapter Five deals with some of these issues from a more political perspective. As time goes on, you will be increasingly well placed to judge, albeit with the benefit of hindsight, whether the points identified by economic theory are indeed a stumbling block to effective implementation in the highly politicized arena of international environmental agreements.

## Summary

- Many environmental issues require responses to be agreed and implemented by more than one country.

- Because countries are sovereign, international environmental agreements need to be self-enforcing.

- Agreements can be made self-enforcing if payoffs are appropriately structured. There are various mechanisms that include: making side-payments, issuing credible threats to retaliate, linking environmental agreements to agreement on other issues and the development of reputation.

- Allowing quota trading in the case of Kyoto might substaintally reduce the cost of implementation.

# 5   Consumption, ethics and well-being
## 5.1 Development and values

Thus far, we have concentrated on the analysis of environmental policy response where economic production is predominant, but in this section, I want to shift attention towards consumption. Many environmentalists have questioned whether consumerism and materialism, and their associated environmental costs are necessary for human well-being. This questioning raises deep issues about the contribution of economic progress to human welfare, which we now explore.

The photograph in Figure 3.11 is from Bhutan, a predominantly Buddhist kingdom with a population of approximately 600,000 distributed over an area the size of Switzerland. Described as a 'gigantic staircase' in the Eastern Himalayas, the country spans three quite different climatic regions and is home to a rich array of flora, birds and other animals. It has a history of cultural exchange with Tibet that can be traced back many centuries and evidence indicates human occupation at least as early as 2,000 BC.

Figure 3.11 Bhuddhist Monks in Bhutan. Commitment to sustainable development and the protection of biodiversity draws deeply on the Buddhist ethos.

Economic development in Bhutan is a recent phenomenon whose beginnings may be marked by the production of the first five-year plan and the construction of the first road, both of which took place in 1961. At the turn of the millennium, the country seemed balanced, perhaps precariously, between economic growth and the preservation of its unique cultural and physical environment. On the one hand, it has expanded its exports of cement, electricity, wood and high quality agricultural products. On the other, it has created a National Environment

Commission to protect the country's cultural identity, and environmental issues have been given a prominent position on the school curriculum. The wearing of national costume in public is mandatory and the number, activities and expenditure of tourists are closely regulated to maximize foreign exchange earnings while protecting the country's artistic and religious heritage. Commitment to sustainable development and the protection of biodiversity draws deeply on the Buddhist ethos, is central to development planning and has resulted in the closure of some projects deemed unsustainable. As the King of Bhutan has put it: 'Gross National Happiness is more important than Gross National Product'.

### Activity 3.6

A simple but interesting and important question is how should Bhutan develop? Write down your thoughts about Bhutan's development policy before reading on. I shall offer you a somewhat personal response by way of comment on this activity in Section 5.4 below. (Hint: Are the pursuit of economic development and the maintenance of cultural identity compatible goals or should the country favour one over the other? Put another way, does income measure what matters to a country or are other indicators of well-being and progress required?)

## 5.2  Can consumption be sustained?

The secular, Western tradition under which economic analysis operates is a philosophical approach known as 'utilitarianism'. An intellectual product of the enlightenment and promoted most notably by Jeremy Bentham in the late 18th century, it proposed that government policy should maximize the 'greatest happiness of the greatest number' (see Figure 3.12). To do this a government needs to know about people's preferences and the modern economic approach has focused on what people are willing to pay (i.e. market prices) as the best behavioural indicators of what their preferences actually are. This is one 'economic' interpretation of utilitarianism and it provides a rationale for estimating economic welfare by calculating the value of consumption. So we have, it would seem, an argument for focusing on consumption or income and asking how sustainable it is.

Although we have already considered sustainability in the context of a particular industry's output (fishing), sustainable consumption raises questions that need to be addressed at the level of national and global economies. As Helm (1998) notes, 'sustainability is a recognition that without intervention, the global environment will not be able to provide a reasonable standard of living for future generations'.

But what do people think about the prospects for sustainability?

On the one side are those who take an optimistic view about the sustainability of current consumption patterns. Bjorn Lomborg (2001), author of *The Sceptical*

**Figure 3.12** Economic analysis operates under the philosophical approach of utilitarianism.

*Environmentalist*, articulates this position with particular passion. Population growth is currently a problem, but one that will decline in future as birth rates fall over the next 50 years. Energy production will see a growing switch from the use of fossil fuels thereby reducing the production of greenhouse gases. (As Lomborg humorously if not persuasively puts it – the stone-age didn't end because people ran out of rocks but rather because they discovered new building materials.) Finally, biodiversity losses will be reduced as scientific knowledge improves and developments in genetic science may allow biodiversity losses to be reversed.

Pessimists challenge these projections. They point out that the USA produces one quarter of the world's $CO_2$ emissions and that there is no sign of switching to renewable fuels on any major scale. Population forecasts for India and China taken together with fuel development plans suggest that greenhouse gas emissions and, consequently, sea levels will continue to rise. And the role of genetic science in fostering biodiversity is doubted (see Chapter 6). We must take steps now, the pessimists argue, to prevent the rise of global conflict and poverty that will otherwise become commonplace in the 21st century.

## Activity 3.7

Are you a policy optimist, pessimist or something in between? Write down your reasons before reading on (you might consider the prospects for global climate change, population growth and biodiversity losses).

## 5.3  What is well-being?

Relatively recently, there has been a shift in emphasis from economic models of consumption towards those that focus on well-being (economists use the term utility). Ecological economists, among others, would say that there are important consequences for the analysis of environmental policy (see also Box 3.2). For one thing, if we take well-being seriously, rather than merely assuming it is strongly and positively related to economic income, then there are reasons to reject conventional measures of economic progress (like Gross National Product, GNP, which measures the monetary values of goods and services produced) in favour of more direct and detailed measures of well-being. In what follows, we shall look at some of these objections before looking at an attempt to construct an alternative indicator of sustainable development.

### Box 3.2 Ecological economics

There are, in environmental economics, two distinct though sometimes overlapping camps. On the one hand, there are neo-classical economists who study environmental issues from the perspectives of efficiency and growth. They have been responsible for developing or applying most of the analytical approaches discussed in this chapter. However, there is a second group, called ecological economists, who emphasise the importance of institutions in determining outcomes. They tend to be concerned with social and environmental consequences of economic activity and use relatively little mathematics in their analyses. In addition, they tend to be on the side of conservation and local self-management rather than big business and the maximization of profit. The empirical sensitivities of ecological economics (particularly when combined with some of the analytical insights of neo-classical environmental economics) can be particularly helpful to policy makers. However, the methodological and theoretical differences between these two camps often hinder the integration of evidence and theory that environmental policy making seems to require.

Three key differences between ecological economics and neo-classical environmental economics are summarized in Table 3.4 below. Firstly, ecological economics tends to give more emphasis to the scale of economic activity and its impact on environments. The neo-classical approach tends to view the environment as a source of inputs and then asks how these inputs might best be exploited. The ecological approach, however, seeks to emphasize other issues such as the links between human well-being and the natural environment as well as our duty to respect and value the environment for its own sake. Related to this is a second issue to do with substitutability of resources, which is related to the optimism/pessimism debate touched on at various points in the text. Neo-classical economics tends to assume that innovation will allow other inputs to be substituted for the environmental resources that are permanently depleted, while ecological economists are more sceptical about this possibility. Finally, and again linked to the previous points, there is a difference between the two schools of environmental economics in the extent to which they believe that human well-being is dependent on the environment (Dawson, 2003). Neo-classical economists tend to see the link being reduced while ecological economists believe human well-being is intimately related to the status of the environment. Table 3.4, adapted from Turner (1994) summarizes some of the differences between these two schools of thought.

Table 3.4   Ecological economics and neo-classical environmental economics compared

|  | Neo-classical environmental economics | Ecological economics |
|---|---|---|
| **Amount of economic activity** | There is modest interest in the relationship between the amount of economic activity and its impact on the global environment | Considerable concern for the impact of economic activity |
| **Substitution between inputs** | Human-made capital can be substituted for natural capital if depleted | It will be difficult to find human-made substitutes for some natural resources |
| **Links between economy and ecological welfare** | Economic growth is increasingly unrelated to the status of the environment | Economic and ecological welfare are closely coupled. |

*Source:* adapted from Turner, 1994.

## Philosophical objections to GNP maximization

As we have noted, economic progress is underpinned by the philosophy of utilitarianism but there are many objections to utilitarianism and they apply, therefore, to conventional approaches to economic progress. For one thing, in seeking the greatest happiness to the greatest number, utilitarians focus on maximization of the total sum of all goods and they are therefore likely to ignore distributive issues. In other words, they can discount a certain amount of misery if the overall level of happiness is at its maximum.

Others oppose utilitarianism for its focus on the present, but there is nothing in the injunction to maximize utility that requires us to confine our attention only to those people alive at present – indeed it seems sensible that we should include the preferences of all future generations (see Chapter One for some of the difficulties in seeking to do this). Nor is there anything inherent in utilitarian theory that prevents us from extending its scope to include concerns about other animals. Even so, ecological economists and ecologists more generally tend to disapprove of utilitarian justifications. Rather they want to emphasize that we should be concerned about future generations and about animals because both have rights that we have a duty to respect.  For instance, a tribal community's right to remain in a particular territory might outweigh the economic benefits from resource exploitation that could be enjoyed if the community moved.

Developing an alternative to utilitarianism that takes such considerations into account, Amartya Sen (1993), a Nobel Laureate in economics, has argued that social planners should be concerned about the capabilities that people have. In Sen's framework, we think of an objective shortlist of valuable actions (e.g. fulfilment of physical needs, abilities to socialize, scope for personal development) to which people might assign their own subjective priorities. Some things that people pursue, like health and longevity, benefit in general from economic progress though others, like integrity, excitement and safety are linked to consumption in much less direct ways. Some cultures, for example, appear capable of providing strong personal identity at a very low level of resource use. From this perspective, being a high-income country could be taken to be a sign that material needs are being met inefficiently! (See Figure 3.13.)

**Figure 3.13** Are happiness and well-being necessarily an end-product of GNP? Two images from the Kumbha Mela festival in India

## Psychology and the value of GNP growth

In 1976, Fred Hirsch produced a book, *Social Limits to Growth*, which drew environmentalists' attention to what he called positional goods. The idea is just that some goods and services are consumed because doing so confers status on the consumer, a point well illustrated by the use of brand names like *Rolls Royce* and *Gucci* as synonyms for luxury and exclusiveness. For such goods, a rise in income levels is self-defeating: as more people have access to such goods, either they lose their value as signifiers of status, or they have to be redesigned to be more expensive so as to maintain their value as indicators of hierarchy.

Turning to evidence concerning the psychological impact of income, we find rather curiously, that increases in income over time make people no happier. We know that poverty and unemployment are major sources of dissatisfaction

(Argyle, 2001) but it may be surprising to know that wealth offers less well-being than we might imagine. The phenomenon, known for some thirty years at least and now beginning to attract widespread attention, is referred to as the Easterlin Paradox for the result is not intuitively obvious and it demands explanation. One possibility is that a relatively high proportion of the goods and services that we consume confer positional benefits and are therefore temporary satisfiers of well-being.

Alleviating poverty is one response to low well-being, but what about these other issues? If we look at other significant economy-related determinants of happiness, we find they include things like the design of work (it needs to be suitably challenging and provide some autonomy), the availability of opportunities to take part in leisure activities (particularly physical ones) or make contributions to civic society (doing voluntary work). Yet, it seems that a rise in the GNP of a country tells us nothing about progress with respect to any of these issues.

## Accounting problems with GNP measurement

Economists have been keen to point out that national income only measures economic welfare, and even then, it has a number of limitations. Firstly, it has long been known that the exclusion from national accounts of unpaid work, like that done in the home or for barter, is unsatisfactory. A much quoted, if dated, example is the fact that marriage to one's housekeeper reduces the national income because the housekeeping services are no longer paid for. In low-income countries the danger is that as more activities are drawn into the economy, increases of income merely reflect the growing monetization of the economy and not improvements to human well-being.

The other point that I want to make concerns the fact that GNP measures flows of income and consumption but not consumer assets. Businesses are required to keep income accounts and statements of assets because one without the other is rather meaningless. It is often very easy to increase income by running down assets and this is very much how non-renewable resources have been used to increase income in previous centuries. This decline in assets is not reflected in the GNP. On the other side, increases in human capital through education, training and experience are also not reflected in GNP (though the costs of producing them are).

○     Provide one example in each case of philosophical, psychological and accounting objections to GNP maximization.

●     Here are three examples. One philosophical (or ethical) objection to income maximization is that GNP ignores questions of distribution. A psychological objection is that rises in income do not correspond to increases in reported happiness. And from an accounting perspective, it is worth noting that work in the home is not currently given a monetary value.

So, if the Bhutanese are doubtful about the value of GNP maximization, then they are not alone. Philosophical views of well-being suggest that humans have needs that monetary wealth, may or may not, satisfy. Psychologists find little evidence of a link between income and happiness, except at very low levels (poverty is a miserable state). And there are many omissions from national income accounts that mean they often do not give us a reasonable picture even of economic welfare. Indeed, given that policy makers are shifting, expanding and reconsidering the ways in which progress and sustainability are measured to reflect some of these concerns, it would seem bizarre to judge Bhutan's progress by GNP growth. (Mike Keefe's cartoon, Figure 3.14, makes a similar point in a light-hearted manner.)

Figure 3.14 Walmart, Everest branch.

There are reasons to do with autonomy and personal freedom that might cause us to be concerned about compulsory cultural protection. Though even then, this is, in effect, what a number of minority linguistic groups within Europe have fought for (e.g. compulsory language teaching). My sense is that a small number of indicators, reflecting the things that *really* matter, should be developed and monitored – indeed we might even ask just how necessary the calculation of national *income* actually is.

## Measuring sustainable development

I want to conclude this chapter by mentioning a debate about the measurement of sustainability that illustrates just how varied results can be depending on how one defines concepts like sustainability and welfare. Towards the end of the 1990s a task force initiated by the World Economics Forum produced a new measure of sustainability which they used to rank over 121 countries around the world. The results of this ranking appear in Figure 3.15:

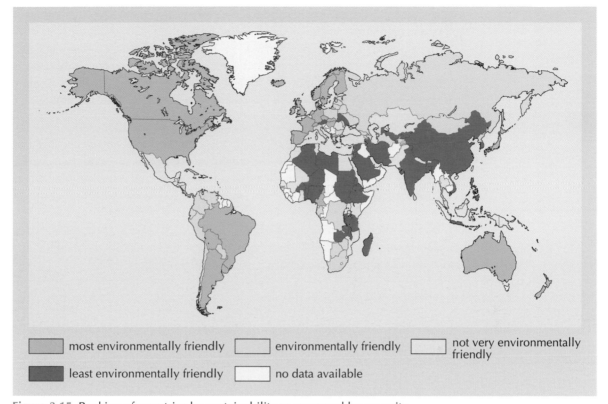

| | | |
|---|---|---|
| ▨ most environmentally friendly | ▨ environmentally friendly | ▢ not very environmentally friendly |
| ▨ least environmentally friendly | ▢ no data available | |

**Figure 3.15** Ranking of countries by sustainability as measured by capacity.

Sustainability is indicated by colour: the greener a country, the more sustainable, and the redder a country the less sustainable judged by the WEF exercise. Basically, the figure shows that high-income countries tend to have more sustainable income and the reason is that they have greatest capacity to tackle environmental problems.

However, Friends of the Earth and *The Ecologist* examined the methodology used and made a number of critical points of which the following are particularly significant:

1  Firstly this index uses 22 equally weighted indicators relating to environmental systems, human vulnerability, social and institutional capacity and global stewardship. It is meant to be an indicator of sustainability but it includes variables such as infant mortality which, though crucial, are only weakly related to the concept of sustainability. Whether child deaths are high or low at any point in time tells us nothing about whether they can reasonably be expected to remain so over the medium to long term.

2    Nearly a third of the indicators (7 out of 22) measure capacity to address environmental problems. Capacity to address environmental problems may be necessary for sustainability but it is not sufficient – we need to know whether this potential is being deployed. For example, one capacity indicator is the number of scientists per head of population. The problem is that we know there is a positive correlation between number of scientists and economic growth, but we don't know about the relationship between the number of scientists and the net impact on the environment. This information is essential if we are to use the variable in our measure of sustainability.

3    Finally there are, inevitably, a number of problems associated with the detailed measurement of variables. In one case, sustainability is measured by the percentage of total forest accredited by the Forest Stewardship Council but this penalizes countries that have large tracts of forest untouched by forestry activities.

To remove these and other problems with the WEF approach to sustainability measurement, The Ecologist and Friends of the Earth modified the index and came up with a ranking of countries as illustrated in Figure 3.16.

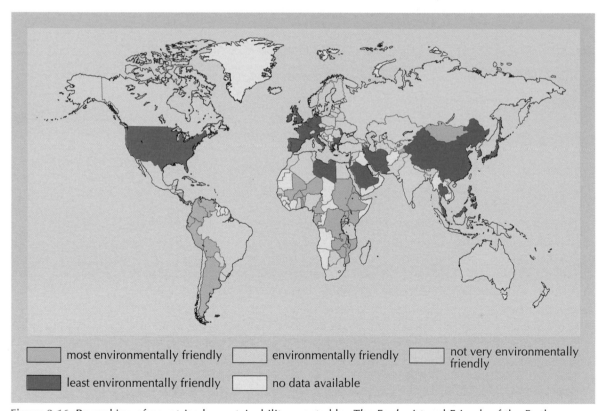

Figure 3.16 Re-ranking of countries by sustainability, created by *The Ecologist* and Friends of the Earth.

As you will see the map of the world is, almost, turned on its head. Those countries that have low-income economies appear to be more sustainable while high-income countries previously judged relatively sustainable have turned red.

The USA is a good example of such a country. In this version of sustainability, it is what one does to the environment that seems to predominate, not the potential to do something about environmental impacts. This exercise, it seems to me, is inconclusive in both versions but it does highlight the need to be clear about what matters and to measure that, and only that. It also suggests, I think, that even sustainability has different facets that we should try to measure separately. Capacity to deal with pollution, for example, is important, but it is far from clear that we should lump it together with measures of actual remedial action. Doing so is a bit like counting up the number of promises people have made without keeping tally of how many they have kept.

## Summary

- Values are key to policy responses to environmental issues because they determine objectives for economic development.
- It is possible to be either pessimistic or optimistic about the scope for sustainable consumption over the next 50 years.
- There are good philosophical reasons to think that utilitarianism (and its economic applications) is deficient as a basis on which to plan environmental policy.
- There is psychological evidence that increasing income does not result in increased happiness.
- Income does have positive impacts on life expectancy and health status though these vary considerably for low-income countries and taper off for high-income countries.
- International comparisons of sustainability exist though the rankings depend crucially on the assumptions made that go into the rankings.

## 6   Conclusion

In this chapter, we surveyed the ways in which economic analysis can help us to understand policy responses to environmental problems. As a result you should be aware that policy choice can depend on uncertainty in a number of ways. When it comes to the management of renewable resources, limiting competitive pressures that threaten depletion of the resource is key. Economic issues raised by pollution are similar and these are often dealt with either by the adoption of environmental standards or the use of economic instruments. Promotion of clean technology illustrates how policy can combine both.

When environmental issues relate to the activity of more than one country, incentives to participate in any agreed environmental policy are crucial as nations are sovereign states. Finally, we observed that when making international comparisons, the particular choice of measures of performance critically determine our ranking of countries. There are good reasons to question the adequacy of using traditional measures of performance such as economic income but we also say that different definitions of sustainable development can give rise to very different views about which countries are doing most to respond to environmental issues, in particular the development of economic activities that are sustainable and sympathetic to a variety of environmental challenges. Time and again we have seen in this chapter the importance of political issues, which shape both our assessment and response to environmental issues. The next chapter takes up this political angle on environmental responses.

# References

Alcamo, J., Shaw, R.W., Hordijk, L. (1990) (eds) *The RAINS Model of Acidification: Science and Strategies in Europe*, Laxenburg, The International Institute for Applied Systems Analysis (IIASA).

Anand, P. (1993) *Foundations of Rational Choice Under Risk*, Oxford, Oxford University Press.

Argyle, M. (2001) *The Psychology of Happiness*, London, Routledge.

Barrett, S. (1997) 'Toward a theory of international environment co-operation' in Carraro, C. and Siniscalco, D. (eds) *New Directions in the Economic Theory of the Environment*, Cambridge, Cambridge University Press.

**Bingham, N. and Blackmore, R. (2003) 'What to do? How risk and uncertainty affect environmental responses' in Morris, R.M. et al. (eds).**

**Bingham, N., Blowers, A.T. and Belshaw, C.D. (eds) (2003) *Contested Environments*, Chichester, John Wiley & Sons/The Open University (Book 3 in this series).**

**Blowers, A.T. and Elliot, D.A. (2003) 'Power in the land: conflicts over energy and the environment' in Bingham, N. et al. (eds).**

**Blowers, A.T. and Smith, S.G. (2003) 'Introducing environmental issues: the environment of an estuary' in Hinchliffe, S.J. et al. (eds).**

**Brandon and Smith, (2003) 'Water' in Morris, R.M. et al. (eds).**

**Castree, N. (2003) 'Uneven development, globalization and environmental change' in Morris, R.M. et al. (eds).**

Chayes, A. and Chayes, A.H. (1991) 'Compliance without enforcement: state regulatory behavior under regulatory treaties', *Negotiation Journal*, vol.7, pp.311–31.

Dawson, G. (2003) 'Environmental sustainability' in *Economics and Economic Change: Macroeconomics*, Dawson, G., Athreye, S., Himmelweit, S. and Sawyer, M. (eds) Milton Keynes, The Open University.

Dales, J. (1968) *Pollution Property and Prices*, Toronto, Toronto University Press.

**Drake, M. and Freeland, J.R (2003) 'Population change and environmental change' in Morris, R.M. et al. (eds).**

Folmer, H. and Gabel, H.L. (2000) *Principles of Environmental and Resource Economics*, Cheltenham, Edward Elgar.

Goodstein, E.S. (2002) *Economics and the Environment*, New York, Wiley.

Helm, D. (1998) 'The assessment: environmental policy - objectives, instruments and institutions', *Oxford Review of Economic Policy*, vol.14, pp.1–19.

**Hinchliffe, S.J., Blowers, A.T. and Freeland, J.R. (eds) (2003)** *Understanding Environmental Issues*, **Chichester, John Wiley & Sons/The Open University (Book 1 in this series).**

Hirsch, F. (1976) *Social Limits to Growth*, Cambridge, Ma., Harvard University Press.

Lomborg, B. (2001) *The Sceptical Environmentalist*, Cambridge, Cambridge University Press.

**Morris, R.M., Freeland, J.R., Hinchliffe, S.J. and Smith, S.G. (eds) (2003)** *Changing Environments*, **Chichester, John Wiley & Sons/The Open University (Book 2 in this series).**

Pommaret, F. (1998) *Bhutan*, Chicago, Passport Books.

Sen, A.K. (1993) 'Capability and wellbeing' in Nussbaum, M. and Sen, A.K. (eds) *The Quality of Life*, Oxford, Oxford University Press.

Tahvonen, O. and Kuuluvainen, J. (2000) 'The economics of natural resource utilization' in Folmer, H. and Gabel H.L. (eds) (2000).

Turner, R.K., Pearce, D. and Bateman, I. (1994) *Environmental Economics*, New York, Harvester Wheatsheaf.

WWF (World Wide Fund for Nature) (2001) *Put Environment at the Heart of European Fisheries Policy*, Brussels, WWF.

# Answers to Activities

### Activity 3.2

There are a number of possibilities you might have mentioned. These include: the imposition of an outright ban on emissions into the river from the factory; a requirement that the polluting factory compensate the salmon farm; the imposition of a tax on pollution into the river; and the requirement that the factory clean its emissions before releasing them into the river.

### Activity 3.4

1 There were a *small number of countries and companies* involved in the production of CFCs making both agreement and enforcement more feasible.

2 The agreement itself could be said to be *fair* in the sense that it required reductions to be made first in countries historically responsible for most of the pollution.

3 Enforcement was promoted by giving low-income countries an *incentive to participate* in the form of financial compensation. High-income countries also had an incentive to participate as the technological development of substitutes opened up *new market opportunities*.

4   Mounting *scientific evidence* about both environmental and health impacts appears to have fostered the agreement and its expansion subsequently.

5   One more issue that might be mentioned is that a non-signatory country could free-ride on the Protocol by enjoying the environmental benefits while importing and using CFCs. By banning the trade of CFCs, the Protocol limited the scope for free-riding in this way.

### Activity 3.5

We can apply the PD game to the problem of over fishing in the following way. Each country has two possible strategies involving low or high levels of fishing effort. The amount of fish caught per annum depends on the combination of strategies that prevails. If both countries allow their fleets to put in high levels of fishing, for example, each will obtain two million tonnes of fish per annum. However, if both countries were to reduce their levels of effort, then each would obtain three million tonnes of fish. Note that either country could obtain four tonnes but only if the other country adopts a low level of fishing effort and accepts a lower catch (one million tonnes). The rationale for applying the PD game to the analysis of fisheries exploitation is that over fishing reduces the fish stock's ability to reproduce itself each year. (Note that you many have different numbers in your boxes, but the principle that an agreement to reduce fishing effort can provide mutual rewards should be a clear outcome in your figure.)

|  |  | country Y | |
|---|---|---|---|
|  |  | low fishing effort | high fishing effort |
| country X | low fishing effort | 3 , 3 | 1 , 4 |
|  | high fishing effort | 4 , 1 | 2 , 2 |

Figure 3.17 Levels of fishing effort as a PD game; yields of fish in millions of tonnes.

# Chapter Four

# Environmental politics: society's capacity for political response

Pieter Leroy and Karin Verhagen

# Contents

# 1    Introduction

Environmental problems represent a major challenge to society. Their societal causes are deeply rooted in a range of activities such as our individual behaviour, consumption patterns and (auto)mobility, and in the structural patterns of our economy and global trade system. The solution of environmental problems therefore requires comprehensive social changes in everything from our individual behaviour right up to the global economic structure. That is what environmental politics is all about – organizing society's response to the environmental challenges it faces.

This chapter deals primarily with environmental politics. Considered as a political response, environmental politics requires concern and mobilization, and particularly social action and political power. These are beyond day-to-day policy making on specific issues; they address the fundamental societal processes causing environmental problems. Although the chapter will deal with some policy issues as well, we shall mainly draw attention to society's capacity to express its environmental concern, to mobilize for social action and to translate the latter into a political response.

Both environmental issues and the changes envisaged to tackle them have social effects: poor air and water quality, for example, clearly affect people's quality of life, but often the solutions also affect their patterns of living and consumption. The effects are unevenly spread among rich and poor, as the former mostly enjoy a better quality of environment and have better opportunities to escape from poor environmental situations. These social effects and their uneven distribution tend to discourage people from endorsing environmental measures. This brings us to another issue dealt with in this chapter: environmental politics is also about mobilizing support not only by presenting scientific evidence but also by gaining **legitimacy** for certain solutions. By legitimacy we mean the extent to which a political system in general or a specific decision in particular is regarded as right, good, effective and fair either by formal institutions or by the general public. To achieve this, **political participation** in the decision-making processes by those affected by the decisions is crucial in gaining support for society's response. Political participation is expressed in the way in which all non-governmental agencies but in particular citizens, either organized or through their political representatives, are involved in political decision making.

*legitimacy*

*political participation*

Our focus, therefore, is on society's capacity to organize its political response to environmental issues. Taking two case studies we shall examine how environmental issues reach the political agenda and the extent to which public support and strategies for action relate to society's openness and the opportunities for participation. The changes from state regulation to market-based strategies of environmental policy will be explained in terms of social and political changes occurring towards the end of the last century in western societies. The final part of the chapter will consider the shifting roles and relationships of the state, market agencies and representatives of civil society in society's response to environmental issues.

# 2    Two environmental issues in their respective political contexts

We start with two cases that are illustrative of the way environmental issues provoke societal action and mobilization. Our first case in particular illustrates the uneven division of power between the actors involved, the mechanisms by which an environmental issue is put on the agenda (or not), and which issue eventually gets sufficient public and political attention to provoke a political response.

## 2.1  The Coto de Doñana disaster

In April 1998, a major environmental accident occurred in Andalusia, southern Spain (see Figure 4.1a). The accident, for reasons that we will discuss, did not get the same public attention as other accidents such as Chernobyl or environmental conflicts such as the Brent Spar, and yet it was a major environmental disaster.

During the night of 25 April, heavily polluted water and sludge flowed into two rivers: Rio Agrio and Rio Guadiamar (Figure 4.1b). The contamination was caused by the collapse of the retaining wall of a basin used for the storage of sludge containing toxic heavy metals from the zinc mining operations of the Swedish company Boliden located in the village of Aznalcóllar (Figures 4.2a and 4.2b). Within a few hours, the contaminated sludge spread over 62 kilometres south from Aznalcóllar along the Guadiamar River route (Figure 4.3a). As you can see from Figure 4.1b, the Guadiamar is one of the main water supplies to the Coto de Doñana National Park.

Immediately a huge controversy arose about the short-term and long-term effects of the accident. Both Boliden and the Spanish political authorities claimed that the contaminated water was stopped near the borders of the national park by a rapidly erected dam (shown in Figure 4.1b). According to the Spanish environmental minister this dam saved the heart of Doñana from an ecological disaster. But environmental non-governmental organizations (NGOs) such as the World Wide Fund for Nature (WWF) claimed that toxic water entered the national park, partly as surface water but, more importantly, via groundwater. The latter could be anticipated to cause huge damage to the ecosystem in the years and decades to come. On 29 April, the director of the Estación Biológica – a centre for monitoring and scientific research established in the park since 1963 – endorsed this diagnosis and stated that toxic water had already entered Doñana or was very likely to in the near future. The park management itself, however, still denied any immediate damage to

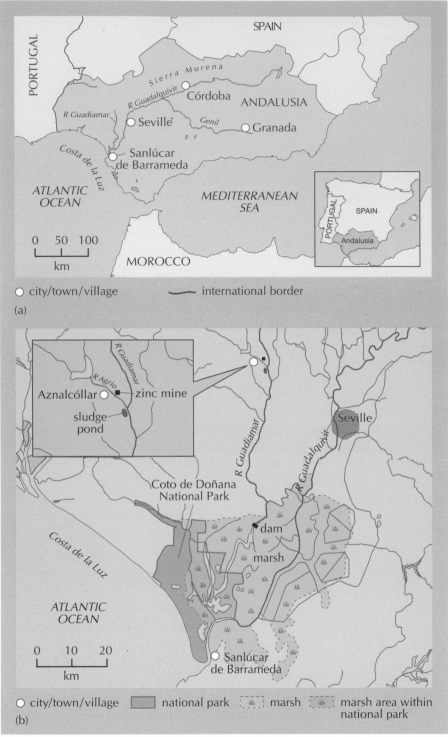

**Figure 4.1** (a) The location of Andalusia in southern Spain; and (b) the location of the Coto de Doñana National Park, to the south of Aznalcóllar and the zinc mine.

the park's natural beauty. Apparently its position was inspired by the fear that a loss of valuable natural habitat would mean a decline in the status of the national park on the one hand, and anticipated possible drops in visitor numbers on the other.

In short, while the accident had a disastrous short-term impact along the river (see Figure 4.3), the controversy mainly centred on the uncertainty about its long-term ecological impacts.

(a)                                          (b)

**Figure 4.2** (a) The zinc mine at Aznalcóllar with the sludge pond in the background, and (b) sludge flowing through the break in the retaining wall.

(a)                                          (b)

**Figure 4.3** (a) The scene five days after the accident, toxic sludge covering the countryside, and (b) a closer view of the environmental degradation, including the heavily polluted water.

## The disaster and its immediate geographical and socioeconomic context

Coto de Doñana, which covers about 50,000 acres, is located to the south of Seville, on the estuary of the River Guadalquivir (see Figure 4.1b). It was established as a national park in 1969, under Franco's dictatorship – an important detail with regard to its perception and image, as we shall see. Its establishment was primarily the result of action from foreign biologists and scientists and from international nature conservation groups, especially WWF, proclaiming its natural value.

Andalusia has a subtropical climate but the national park itself has particular biological conditions and its own microclimate due to its location in the Guadalquivir delta and the moderating influence of the nearby Atlantic Ocean. It contains very special flora and fauna (Figure 4.4) with, for example, 361 species of birds and more than 820 species and subspecies of vascular plants. It serves as a stop-off for migrating birds between Europe and Africa, and it is one of Europe's most important bird reserves. Moreover, it represents an important link in a network of natural parks all over Europe and has a crucial role in the preservation of Europe's biodiversity. About 400,000 tourists per year visit the Coto de Doñana, mainly in combination with visits to the historic cities of Seville, Córdoba and Granada (shown Figure 4.1a).

Figure 4.4 The biodiversity and natural beauty of the Coto de Doñana region, an essential link in Europe's network of natural parks.

Apart from the historic cities mentioned, Andalusia is not the wealthiest region of Spain, and the area surrounding the Coto de Doñana National Park in particular is relatively poor. Some local citizens even regard the national park as an obstacle to further economic development, as it covers a large area and impedes traffic and transport between Andalusia's different economic regions. Economically the area is primarily dependent on agriculture (rice, cotton, olives and citrus fruit) and cattle breeding, whereas industrial activities on average are limited. However, the huge Boliden mining plant at Aznalcóllar, which dominates the village and its landscape, lies about forty kilometres to the north of the park's borders. Further to the north and north-west, there are a series of traditional mining industries, some of which even date from Roman times. Apart from their impact on the landscape, these industries also have a long history of environmental degradation in terms of air and water pollution. The latter is reflected in the name of one of the rivers nearby, the Riò Tinto, literally 'red-coloured river'.

## Activity 4.1

Why do you think that people in the local region have somewhat mixed feelings about the national park?

### Comment

First, the park was established in 1969 under Franco's dictatorship and many people still associate the park with the dictatorship and the oppression of that time. Second, since the park is a strictly protected area, the development of economic activities in it or in its direct surroundings is prohibited, thereby hindering further economic growth. Local people perceive this situation as disadvantageous because of their relatively poor economic situation.

Aznalcóllar itself is a typical example of a so-called 'one-industry community' – a village or community that economically depends heavily on a single (industrial) activity, in this case Boliden mining. These one-industry communities are very often 'peripheral communities' located in somewhat remote areas, at a distance from a nation's political and economic centres (Blowers and Leroy, 1994).

Peripheral communities in relation to the nuclear industry are discussed in **Blowers and Elliot (2003)**.

A typical feature of one-industry communities is the symbiosis between the community and the industry, not only in economic terms, but also politically, socially and culturally. The economic dependency can be measured in terms of employment, local income, consumption and so on. As the local community is highly aware of its economic dependency, it embraces the local industry as the very basis of its sustenance and it organizes its social and political life accordingly. In many one-industry communities, therefore, industry representatives are mayors or aldermen and the industry sponsors the local sports clubs and organizes various cultural events. Even local education opportunities may depend heavily upon the needs of the industry.

### Activity 4.2

Can you think of any examples of one-industry communities from your region or country? Do they reflect all the characteristics mentioned above?

Clearly, the symbiosis between the community and the one local industry makes the former vulnerable to the ups and downs of the latter. In our case, the economic dependency to a large extent explains the controversy in the immediate aftermath of the disaster between the citizens of Aznalcóllar and their local political representatives, on the one hand, and some international NGOs on the other. International environmental NGOs engaged in nature conservation, such as WWF, expressed their deep concern about the compatibility of having mining industries in the surroundings of the national park. Indeed they claimed that the mining industry should be closed down. But within the region, let alone in Aznalcóllar, nobody echoed that claim. On the contrary, within days of the Aznalcóllar disaster, workers of the Boliden company and citizens of the village organized a demonstration, loudly expressing their protest against a possible closing of the mining company. Significantly, one classic actor in this kind of environmental issue is missing here: there is hardly any local environmental action group in the region.

## The disaster and its political context

Apart from the controversy over the long-term impact of the disaster, another controversy unravelled in the days and weeks following. Local and regional authorities had a series of disagreements with the national government over their respective responsibilities. To understand those quarrels, we need to look briefly at the political context. Following the end of Franco's dictatorship in 1975, Spain not only democratized, but also went through a major devolution of competencies to the regions. Devolution meant that competencies previously in the hands of central state bodies in Madrid, gradually became the responsibility of regional authorities, in this case the regional authority in Seville.

Devolution, however, does not always lead to a clear division of responsibilities. Particularly when competencies compete, it may result in quite complicated political situations. In our case, the Andalusian regional government was responsible for the control of the mining industry and therefore was held responsible for its failing, whereas the Boliden plant's permit itself was issued years before by the national authorities. So the question arose as to who was to be held responsible for the disaster. Another quarrel arose regarding the geographical division of responsibilities: the regional government in Seville is responsible for the protection of the so-called 'Natural Park' that surrounds the Coto de Doñana National Park (see Figure 4.5). The latter, however, is still the responsibility of the national authorities in Madrid. This complicated and controversial division of competencies prevented a common political response to the problems caused by the disaster. Existing patterns of political interrelations

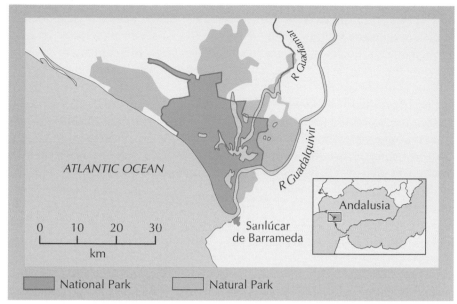

**Figure 4.5**  Coto de Doñana National Park and Natural Park.
*Source*: adapted from Sunyer and Manteiga, 1998, p.86.

and administrative responsibilities can thus facilitate or hinder society's capacity to respond to an environmental issue.

Yet the regional authorities in Seville realized that the disaster possibly created a great opportunity. It unexpectedly offered them a chance to implement a plan that they had already prepared: the establishment of a 'green corridor' along the Rio Guadiamar, linking the natural beauty of Coto de Doñana to that of the Sierra Morena, in the north of Doñana (see Figures 4.5 and 4.1a). This green corridor would be part of the European Natura 2000 network of nature reserves and buffer zones, connecting them into one huge 'green infrastructure' and thereby contributing to Europe's policy on biodiversity. For the regional authorities in Andalusia the plan had quite a lot of political prestige: it could demonstrate their determination to develop their own nature conservation policies and it would be funded by the EU rather than by the national government. Nevertheless, this plan, initially launched some years earlier, had not gained much attention either from the local media or from local people. Nature conservation was not their political priority and so the plan did not provoke any action.

See **Freeland (2003)** for a discussion of the role of wildlife corridors.

But the plan now gained attention and was, albeit for different reasons, supported by almost all parties involved. Regional authorities saw it as an opportunity to realize their policies on nature protection, thereby trying to overcome the negative effects of the disaster and giving it a positive perspective. And the disaster itself made other groups change their minds. So far, farmers had opposed the green corridor mainly because it would cause losses of agricultural land. Now, however, as their grounds along the river were heavily damaged by the disaster, these farmers were not unwilling to have them expropriated by the regional government. The tourist sector in the region had so far not made too much profit from the park, as it imposed rather strict rules and conditions on tourists wanting to visit its beauty. Now the tourist sector anticipated that the green corridor could offer new opportunities for environmentally friendly tourism, for example the organization of horse riding, mountain biking and walking tours along the tracks in the area. The park authorities of both the National Park and the Natural Park could hardly oppose a plan that aimed at making them part of a more robust natural infrastructure. International nature conservation organizations found themselves in a similar position. And both the Boliden mining company and the citizens of Aznalcóllar – as far as they were to be involved in the discussion at all – could continue their economic activities unaffected by the green corridor.

As a consequence, within a year of the disaster, it was overshadowed by a plan that was supported by almost everybody. As no one within the region had any interest in dramatizing the disaster, it now seemed to be played down. And yet it is important to realize that neither the long-term environmental damage nor the sustaining environmental threats were taken away. The political upheaval faded away, but the environmental issue itself was not solved. The latter, apparently, was beyond the region's and its people's **political capacity**. The concept of political capacity indicates the extent to which a society through its political system is successful in handling and resolving the problems it faces. Its success depends on the resources available and the extent to which it gathers support and legitimacy from the general public, parliament, pressure groups and other agencies.

political capacity

### Activity 4.3

(a) In what ways did the Coto de Doñana disaster lead to social and political action?

(b) Which of the following agencies were relatively powerful, and which were relatively powerless? And why?

international NGOs

the citizens of Aznalcóllar

local and regional authorities

### Comment

(a) The disaster led to societal action against the possible closing of the mine: citizens of the village and workers of the Boliden company co-organized a demonstration. Regional authorities in Seville took political action: they finally got the opportunity to establish a green corridor. As a consequence of the disaster some groups who at first opposed the green corridor changed their minds. Among them particularly were local farmers, who were affected by the disaster.

(b) International NGOs made people in other countries aware of the magnitude of the disaster. However, they were not able to influence the opinions and reactions in the affected region itself. In reality, they were relatively powerless.

The citizens of Aznalcóllar and workers of the Boliden mining company depend heavily upon that company. Therefore their economic situation was very vulnerable. Although relatively powerless, by their protest they anticipated a possible closing of the mine, which was campaigned for by international NGOs but never seriously considered by regional authorities.

Because of the disaster local and regional authorities implemented a plan for a green corridor, a plan that had hitherto lacked support. They were relatively powerful but nevertheless had to recognize the need for a solution broadly acceptable to all interests.

## Some lessons and conclusions

The Coto de Doñana case provides some insights into society's capacity to respond to environmental issues. The following are some of them:

- An environmental disaster, even when dramatic and overwhelming in itself, apparently is not sufficient to provoke societal action, let alone provoke a political response. Compared to the magnitude of the Coto de Doñana disaster, the Brent Spar case (see **Smith, 2003**) was a minor incident in ecological terms, and yet the latter gained much more social and political attention. We can therefore conclude that the attention an environmental problem gets and the response it provokes does not relate so much to its ecological magnitude, but depends largely upon social and political processes. The latter in turn depend upon the power relations and capacities of the actors involved.

- In this particular case, the economic dependency of the community, and to a lesser extent the region as a whole, explains why those affected by the disaster, both local citizens and farmers, did not mobilize against the mining industry. They were rather powerless, and drawing attention to the disaster could even further undermine their already weak economic position. We therefore tend to conclude that an environmental disaster does not alter the pre-existing power balance. On the contrary, it seems as if environmental burdens tend to reinforce existing political inequalities, as people and communities from underdeveloped areas very often are unable to mobilize a societal response to environmental threats. In this particular case even nature conservation groups from abroad could not initiate demonstrations by local citizens, nor in any other way empower them.

- This particular disaster led to a wide acceptance of a plan that until that point had not been supported. In that sense the disaster had mobilized the societal and political energy to build up consensus and to endorse a political response that might bring some benefit to the region.

The Coto de Doñana case illustrates the way environmental issues provoke (or otherwise) societal mobilization and political action for specific reasons. The case illustrated the uneven division of power among the actors involved and the mechanisms by which an environmental issue is, or is not, put on the societal and political agenda. Our second case will provide some further evidence of how environmental issues are put on the agenda, how they provoke societal and political action, and what the actual outcome of all that may be.

## 2.2 Constructing a controversial railway line: the Betuwe railway

The second case, the Betuwe railway, originates from the Netherlands. It illustrates the societal and political response to environmental threats under quite different circumstances from those in the Coto de Doñana case. From the late 1980s onwards Dutch economic politics was dominated by the slogan *Nederland Distributieland*, literally 'the Netherlands: distribution country'. The slogan reflects the idea that the strength of the Dutch economy largely depends upon its capacity to distribute goods within Europe and on a global scale. Some transport hubs, particularly Rotterdam harbour and Schiphol airport, were said to be crucial for the further development of this, and main road and train connections should link these economic pivots to the 'hinterland' – hence the initiatives for the extension of Schiphol airport, the building of a new industrial area at Rotterdam, the high-speed train connection to Brussels, London and Paris, and the construction of the so-called 'Betuweroute' or Betuwe railway, the subject of our case study here.

The Betuwe railway project (see Figure 4.6) involved the construction of a brand new railway line connecting Rotterdam harbour with the German Ruhr region, the two main economic areas in their respective countries and two regions of great economic interest in north-western continental Europe.

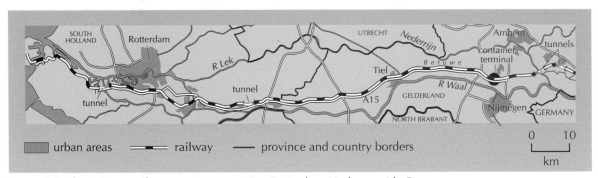

**Figure 4.6** The Betuwe railway route, connecting Rotterdam Harbour with Germany.

## Political context: protest and support

From its first announcement in 1989 the Betuwe railway initiative provoked protest from citizens, local authorities and local environmental and conservation groups. The line would cross a densely populated area, and although its route would parallel existing infrastructure, such as highways and railroads, it would also cross villages and communities. On top of that it would cause considerable noise pollution and further affect the landscape along the whole route when crossing rivers and areas of natural beauty. Hence a large protest movement arose.

At first, though, the Dutch environmental movement at national level did not participate in the protest. On the contrary, it initially supported the idea of the railway connection, on the basis that it would provide a substitute for part of the road traffic and thereby diminish the environmental damage per tonne or per kilometre. If railway traffic could indeed substitute for road traffic, it would contribute to a reduction in both the use of resources and the emission of pollution. From the early 1990s onwards, though, there was growing scientific evidence that the Betuwe railway would not contribute to substitution from one traffic mode to another, but rather would further the growth of traffic in general. In addition, its overall contribution to possible economies of energy use and to a decrease in pollution did not appear to be substantial. By this time the environmental movement had already changed its position. In the late 1990s even national organizations had to admit that there was hardly any environmental advantage involved. But by then the decision was already made and the construction was underway.

During the 1990s the Betuwe railway initiative was *the* environmental issue par excellence in the Netherlands. It seemed as if everybody was mobilized: local authorities, fired up by their residents, regarded the project as a major threat to their local communities and their local environments; national and regional environmental organizations, united in a coalition against the Betuwe railway, pleaded for ecologically sound transport modes; and local and provincial authorities, environmental organizations and engineering agencies came up with a series of alternatives, mainly aimed at decreasing local noise pollution and landscape damage. Economic benefits and environmental costs were the main issues in what evolved into a national debate. Almost every week the issue was on the parliamentary 'question time' agenda, provoking sharp debates and controversies between the political majority and the opposition. But by the time the national debate reached its peak, the decision to build the railway line had already been taken by both the Dutch Ministry of Transport and the Dutch National Railway Company. Therefore, the debate was not actually on the principle of the new railway itself nor on its economic necessity, purpose or environmental side-effects. Instead, the debate on the railway focused on its route, its construction, its (underground) crossings of villages and on compensatory measures (see Figure 4.7). In short, the debate was on a series of mitigating measures that could help to diminish harm and damage to both people and the environment.

Authorities at the national level - the minister, the government, the Railway Company and the parliamentary majority – jointly succeeded in overriding any debate on the principle of building the railway. Time and again they echoed the argument brought forward by its initiators – the crucial importance of the railway for the Dutch economy – even when hard evidence for that argument failed. Since all their protests were overridden, citizens' groups, local authorities and

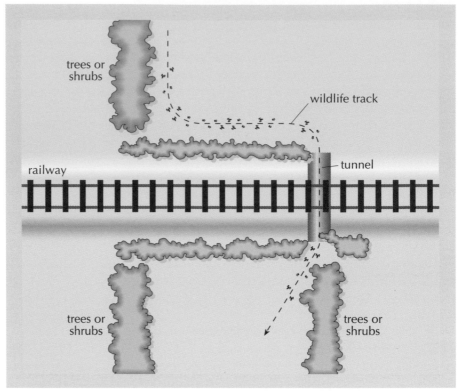

**Figure 4.7** Example of a mitigating measure: a wildlife track.
*Source*: Jongman, 1995, p.104.

environmental organizations were forced to focus their efforts on a series of legal procedures related to such things as building permits, spatial planning, environmental permits and environmental impact assessments. We shall return to the topic of such 'opportunities for participation' later on, when we discuss their establishment in the 1970s and 1980s.

## Protest and influence

In this particular case the legal procedures mentioned above were the only opportunity to give voice to the societal protest and to exercise some power to counter, hinder and postpone a series of decisions on the Betuwe railway. In some local cases the protesters even succeeded in influencing and altering these decisions. The series of protests at local level, the series of legal procedures and the delay caused by them forced the minister and the National Railway Company to agree to a series of mitigating measures (see Figures 4.7 and 4.8) in almost every village and at every crossroads. In doing so, they succeeded in largely neutralizing the protest. At the same time, however, they hugely increased the total cost of the railway: from the initial €2.3 billion in the early 1990s the cost doubled to more than €4.6 billion in the late 1990s, and with its construction underway (late 2002) even the government itself anticipated a further increase in the total cost, most probably up to €6 billion.

The term 'mitigation' has different interpretations here and in Chapters Five and Six.

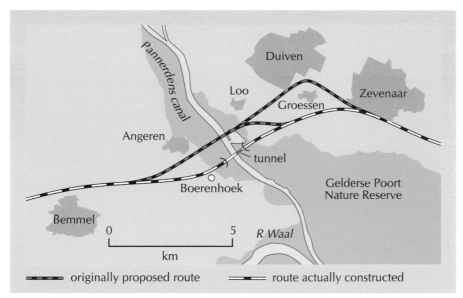

**Figure 4.8** Betuwe railway crosses close to the nature reserve, *Gelderse Poort*. As a result of protests, the originally planned route was abandoned and a new route to the south, including a very expensive tunnel, was built.
*Source*: Pestman, 2001, p.106.

Given this high price and the reluctance of German counterparts to connect the Betuwe railway to the German railway network, it is now widely accepted that the Betuwe railway will never be economically profitable. The initiative goes beyond the traditional economy–ecology dispute, as its ecological balance will still be negative even though it has been built with the most up-to-date techniques to prevent environmental harm. Local and national environmental protests against the project did not succeed in changing the initial plans. Protest was, to a large extent, bought off by mitigating measures that the Dutch public budget apparently could afford to provide.

## Summary

Comparing the Betuwe railway case with the Coto de Doñana case we can draw some conclusions about society's capacity for action and political response, and about power and uneven power balances.

- The environmental threat in the Betuwe railway case provoked huge societal mobilization and protest. Whereas in the Coto de Doñana case people were almost unable to mobilize, mainly since they were economically dependent and thereby politically powerless, in the Dutch case we witness the emergence of a wide protest movement at both local and national level.
- Although the capacity for societal mobilization and action clearly differs from what we have seen in Coto de Doñana, we also see some similarities. Despite its capacity for mobilization, Dutch environmental protest lacked the power to make the authorities withdraw their initial plan or to make them change it

substantially. In other words, the capacity to mobilize did not bring about a substantial change of power relations. On the contrary, the outcome of the process reflects the uneven balance of power.

- This uneven balance of power has not been influenced by the increasing amount of evidence on the non-profitability of the huge investments in the railway on the one hand, and on its non-contribution to further environmental quality on the other. Scientific evidence seems not to be the decisive factor in society's response to environmental issues. Both social and political factors prove to be of greater importance.

- The Coto de Doñana case has so far resulted in a widely supported political initiative in which nature conservation and tourism are to be reconciled, giving the region a perspective on economic development while preserving its natural beauty. It is too early to assess the implementation and final effects of this initiative. The initiative certainly does not solve the issue of environmental threat, but at least it seems to have gained wide acceptance and legitimacy, and has thereby clearly increased the region's political capacity and power. In the Dutch case, however, despite the support for the Betuwe railway by the government, parliamentary majority and captains of industry, the project itself will probably never be fully accepted by the general public. We might even argue that the overwhelming political power with which the project has been pressed through has increased people's mistrust of constitutional politics.

# 3    The emergence of environmental issues as societal and political issues

So far we have looked at the way particular environmental issues – a pollution disaster in one case, a new railway in the other – have led (or not) to social action and to political response. We discovered some social and political factors and mechanisms by which environmental issues gain attention and lead to action (or otherwise). In this section we widen the scope. First we look back at the way in which, from the late 1960s and early 1970s onwards, environmental issues at large were put on the public agenda and led to political responses, at least in most western countries. Then we shall return to our two cases, paying attention to some further similarities and dissimilarities between them. Once again we shall see that social action, political response, power and legitimacy are the crucial factors in the public and political agenda setting of environmental issues.

## 3.1  Back to the early days of environmental concern

Looking back at the way environmental issues have been brought to public attention from the late 1960s and early 1970s onwards, three factors seem to have been decisive: a series of alarming publications, some dramatic disasters and some specific images. They led to an increase in environmental concern among the general public, which is reflected in a series of public opinion polls, public inquiries and other sources of evidence, and can also be seen in the growing number of press articles and television programmes devoted to the subject. To be effective this concern needs to be transformed into societal action, for example by the environmental movement campaigning for a political response. As we shall see, though, the success of the environmental movement does not only depend upon its strategic capacity, but also on 'political opportunities'. The latter are discussed in section 4.

The increase in public concern was, first, provoked by some disturbing publications. Two of the most widely known are *Silent Spring* written by Rachel Carson in 1962, which sounded the alarm over the accumulation of insecticides, DDT in particular, in the food chain (Carson, 1962), and *Limits to Growth*, published in 1972 and warning about the increasing pollution and rapid depletion of natural resources, which had a tremendous impact upon public opinion and political concern (Meadows et al., 1972). We could easily list a dozen other books and brochures dating from that period and covering issues such as pollution and its impact on public health, the environmental costs of our economic growth, the devastating role of modern technology and the depletion of major natural reserves. The feeling of urgency about the environment that they contributed to was reinforced by a series of environmental accidents. In the UK, for example, the sinking of the *Torrey Canyon* in 1966, the first major accident with a huge oil tanker in the Channel, proved catalytic in increasing public concern. In addition, some images also contributed to the spreading of environmental concern. From December 1968 onwards, as the US space programme continued, all papers and magazines carried thought-provoking pictures of planet Earth, wrapped in blue and white and seen from the moon for the first time ever, apparently unique and, above all, very fragile. Today's generation is used to these images, but quite rightly they have been labelled 'the most effective environmental message of the [twentieth] century'.

These publications, accidents of different sorts and sensitizing images were the most powerful motors of environmental concern in the early 1970s. Not only did they bring the environment onto the public agenda, they also led to societal action such as the emergence of the environmental movement. Throughout western countries environmental action groups, be it at local or national level, were established in the early 1970s. Their emergence and the increasing public environmental concern forced public authorities to respond to the issue, with a series of legislative initiatives leading to environmental policies. We will discuss these developments in more detail in section 5.

Despite these developments, however, environmental accidents continued to be very important generators of political response both at national and European level. The gradual establishment of policies on, for instance, industrial risks, toxic waste management, food safety and others is clearly related to environmental accidents such as those at Seveso, Italy (1976) and Basel, Switzerland (1986) – both explosions in chemical plants which raised awareness of hazardous activities and risks – and to crises such as bovine spongiform encephalopathy (BSE) and foot and mouth disease (in the 1990s). The point is also illustrated by the fact that, although German environmental policies were developed from the early 1970s, it was not until the Chernobyl accident in 1986 that a German ministry of the environment was established.

### Activity 4.4

Three main factors have been mentioned that were decisive for the attention that environmental issues gained from the late 1960s onwards.

(a) Which three factors were mentioned?

(b) Can you give some examples?

(c) Why, in your opinion, did these factors cause environmental concern?

The answer to this activity is given at the end of the chapter.

## 3.2  Environmental concern and agenda setting in contemporary cases

Today, environmental issues are widely acknowledged. Yet this environmental concern is not evenly spread among people and among nations, as we can learn from evidence gathered by opinion polls. There are a series of variables that are used to indicate environmental concern and its spread and evolution. One such variable is the varying dates of establishment of environmental organizations in different countries (see Table 4.1).

Table 4.1    Establishment of environmental organizations in different European countries

| | Germany | UK | Netherlands | Italy | Spain | Greece |
|---|---|---|---|---|---|---|
| 1971 | | Friends of the Earth | | | | |
| 1972 | Federation of Citizens' Groups for Environmental Protection | | Stichting Natuur en Milieu (Foundation for Nature and the Environment) | | | |
| 1977 | | Greenpeace | | Amici de la Terra (Friends of the Earth) | | |
| 1979 | | | | Lega per l'Ambiente | | |
| (late) 1980s | | | | | Several Spanish environmental organizations | Several Greek environmental organizations |

*Source*: Weale et al., 2000, pp.258–65.

Our two case studies also illustrate uneven levels of environmental concern and, therefore, differences in agenda setting. First the two case studies provide the contexts for drawing out differences in the process of agenda setting.

**Agenda setting** is a metaphor used to refer to the way issues varying from    agenda setting
unemployment and migration to public health and public safety are brought to the attention of the general public and those with political responsibility (Bachrach and Baratz, 1970; Kingdon, 1984). An issue is said to be on the 'public agenda' when the general public is concerned about it, discusses it, and initiates social action on it, thereby demonstrating its general interest and urging politicians to act on it. The items discussed in newspapers, television programmes and letters to the editor provide useful indicators of what is on the public agenda. An issue is on the 'political agenda' when politicians indeed recognize the urgency of it and initiate political responses, ranging from parliamentary questions to legislative measures. It should be clear, however, that neither public nor political agenda setting are 'automatic' processes and the political agenda does not mirror the public one or vice versa. On the contrary, these are social and political processes, having their own rationales. Therefore they need to be analysed and understood in terms of social action, political support, power and legitimacy.

Our two cases illustrate the way in which an issue reaches the public and political agenda, the decisive factors within these processes and some striking differences in agenda setting. The Coto de Doñana case represents a particular process of agenda setting. No local citizen group or local environmental organization took action to put the environmental issue onto the public agenda. On the contrary, the citizens of Aznalcóllar did everything within their power to minimize the event, specifically to prevent the closing of the mining company on which they largely depended. The greatest worry for the management of the national park was a possible decrease in the status of the park and the number of its visitors. So instead of these two local actors taking action and initiating protest, for quite different reasons they in fact had a similar interest in playing down the events. The governmental bodies did not want to dramatize the events either; after all, the accident revealed the shortcomings of their earlier policies. Farmers affected by the incident were rapidly compensated and the local tourist industry was given new perspectives. So the only actors sounding the alarm were some international environmental organizations, but they hardly got any support and legitimacy from the region itself, only succeeding in bringing the accident to the attention of the international press.

The Coto de Doñana case illustrates an interesting feature of the agenda setting of environmental problems: the physical impact of an environmental problem in terms of pollution, depletion, etc., even when backed by scientific evidence, is not the decisive factor in determining the public and political attention it gets. What really matters is the way an environmental issue is identified and presented, and the way it is socially defined and endorsed (or not) by public action (see **Smith, 2003**, on making environmental news). Environmental groups and organizations in particular play a crucial role in this specific stage of agenda setting. Over the years, Greenpeace for example has delivered some masterpieces in terms of presenting environmental issues and bringing them to the attention of a large audience. It has been particularly successful in reversing the 'out of sight, out of mind' situations by bringing far away issues such as the dumping of nuclear waste, the endangering of exotic species and the threats to the Antarctic to our TV screens – literally into our living rooms (see Hannigan, 1995).

### Activity 4.5

Despite overwhelming evidence, the Coto de Doñana disaster did not provoke public action urging new environmental measures. On the contrary, the action was aimed at playing down the significance of the events. Therefore, it seems to illustrate the 'un-making' of environmental news.

How could you explain this in reference to the socio-economic situation of the region?

### Comment

The essence of the answer from this chapter's point of view of course is that under certain circumstances of relative powerlessness and economic dependency, environmental problems seem to be a luxury.

The Betuwe railway case reflects some different mechanisms of agenda setting. The captains of industry of the Rotterdam harbour and the Dutch trade and transport sector launched the initiative, endorsed by the slogan 'the Netherlands: distribution country'. They succeeded in bringing the initiative onto the political agenda and they convinced politicians of the necessity for a new railway connection. The views of the general public were overruled by the decision of the Ministry of Transport (endorsed by a parliamentary majority) and the Dutch National Railway Company. And yet that decision claimed to have taken into account environmental issues: the railway was presented as a reconciliation of economic and ecological interests.

The national environmental movement at first tended to agree with this. But local groups did not, and argued for their own 'environmental aspects', largely local in character. Not only did they successfully mobilize citizens all along the route and in the country as a whole, thereby putting the issue explicitly on the public agenda, but they also successfully convinced local authorities to endorse their protest. The national environmental movement reconsidered the issue and backed the protest. Citizens and local and national groups jointly then used legal procedures to empower their protest and bring the issue onto the political agenda. Whereas the agenda setting in the Coto de Doñana case suggests power relations that could occur in similar circumstances elsewhere, the pattern of social mobilization in the Betuwe railway case provides a good example for a wide variety of other environmental issues and conflicts.

Given the outcome, the assessment of the success of the protest in the Betuwe railway case is rather paradoxical. On the one hand the protest succeeded in achieving mitigating measures all along the track and getting compensation for losses of natural resources. The protest also profited from the scientific uncertainties on both the economic and the ecological aspects: by initiating legal procedures it succeeded in referring items back for further research, while at the same time building upon the contributions by scientists also opposing the railway. On the other hand, though, the success of all these efforts is relative, as the protestors did not succeed in having the railway construction plan withdrawn and had to be content with a series of local compensatory measures. In other words, and compared to the Coto de Doñana case, despite the huge differences in the magnitude of the environmental protest, its actual success does not seem to be that diffcrent, and could even be vicwed as disproportionate to the efforts. This rather paradoxical conclusion raises questions that we address below.

## Summary

The two cases provide quite interesting insights into the process of public and political agenda setting of environmental problems. Restricting ourselves to looking at the local actors involved, their role in the agenda setting and the actual outcome of the whole process, we have two quite different cases here:

- In the Spanish case local citizens, the company and local governmental bodies all deliberately minimize the problem. Their position is quite understandable, as possible environmental measures risk further undermining their position. They appear to be a de facto coalition that attempts to de-dramatize the event, and in that are quite successful. And paradoxically, at the end of the day they seem to have put a new perspective on the disaster.

- In the Dutch case, local citizens and local environmental groups urge their local governments to forge a strong protest coalition. Their position is quite understandable as well, as none of them can anticipate any benefit from a railway line passing by – the railway, on the contrary, is expected to decrease their quality of life. Their coalition of protest successfully hinders and postpones the construction of the railway line, and secures a series of mitigating measures. Despite their political power, however, the protest did not succeed in having the construction stopped.

# 4   Environmental issues, political power and the political opportunity structure

Our analysis of the two cases so far has focused on the process of public and political agenda setting. We considered the role of social action and protest as crucial elements in this process, and we emphasized the importance of the way an issue is presented and brought to attention. All these factors determine whether an issue is successfully put onto the political agenda and provokes the political responses required. In other words, and in terms of agenda setting, social action seems a necessary though not sufficient condition for bringing an issue to political attention, whereas political attention seems a necessary, but again not sufficient, condition for political response. In this respect the two cases showed similarities as well as differences, with regard both to the mobilization and agenda setting efforts and to the outcome of the process. These differences, we suggested, relate to the political power that the actors involved were able to mobilize.

Thus far we have conceived 'power' as mainly the result of an actor's capacity to mobilize resources. By 'resources' we mean such things as support from the general public, the scientific evidence (or, in some cases, uncertainty), support from environmental and other groups and organizations, media attention and money. It is clear that the political power of an actor largely depends upon his or her capacities to mobilize these resources. When actors lack many, or in some cases just a few, of these resources they will find themselves in a position of relative political powerlessness.

Political power, however, not only depends upon someone's capacities to mobilize, but also on the opportunities the political structure or context provides (Kriesi et al., 1995). Certain characteristics of the political structure or context may indeed impede the mobilization of resources or, in other cases, further it. Using this so-called 'political opportunity approach' we will look at two major features of the political context, namely its 'openness' on the one hand and 'environmental concern' (as discussed earlier) on the other. We will conclude by looking at environmental justice and injustice, as this issue relates to the different power positions of people affected by environmental threats.

## 4.1  The openness of a political system

political openness
Openness of systems, and perceptions of openness, are also discussed in Chapter One.

The **openness of a political system** essentially refers to the 'sensitivity' of a political system to new societal demands and questions, and the way it deals with them. We can, of course, very roughly distinguish between authoritarian and democratic regimes. But even if we restrict ourselves to the latter, there are great differences. Let us consider just three characteristics. First, in some political systems ethnic, religious or linguistic cleavages traditionally dominate any political debate. Such a political system does not leave much space for newly emerging interests, be they environmental or other issues. In Belgium, for instance, linguistic quarrels dominated the scene from the 1960s up to the 1990s, resulting in a delay in, among other things, the development of environmental policies. There was simply no place left on the political agenda.

Second, and similarly, the electoral system of a state may also hinder or further the emergence of new political parties. In general, a constituency-based system in which the winner takes all makes it difficult for new political groups to achieve representation. Therefore, the position of the Green party in the UK – with a constituency-based electoral system – is quite different from that of a Green party in Germany, Belgium or the Netherlands. The latter countries, and in particular their electoral systems, offer more opportunities for a newly emerging political movement to come into the political arena. In other words: the mobilization of resources is more rewarding in the latter cases than in the former.

Third, the concept of openness also refers to the way traditional political elites behave vis-à-vis newly emerging political trends: the former can, in reaction to the latter, aim at either cooperation or at confrontation, use either an inclusive or an exclusive strategy and so on. It is clear that the more a leading political elite behaves in an unfriendly or antagonistic way, the more difficulties a newly emerging movement will face in acquiring the political position for which it aims.

## Activity 4.6

In the above, three characteristics of a political system have been mentioned that determine whether a new issue – in our case an environmental problem – gains attention or not. What are these three factors?

## Comment

The three characteristics are:

- Predominant cleavages in the political system that bias political debates and associated issues, including the blocking of particular issues from accessing the political agenda.
- The election system can be favourable for newly emerging issues and interests, but can on the contrary discourage and hinder them from being brought forward. We contrasted a constituency-based election system such as that in the UK, leading to a 'winner takes all' situation, with an election system that induces a wider political pluralism, such as in Germany, Belgium and the Netherlands. We looked at the consequences of these different systems for the possible rise of a Green party and its entry into Parliament.
- The predominant style and behaviour of traditional political elites vis-à-vis newly emerging political movements. Different styles and types of behaviour can be cooperative or confrontational, inclusive or exclusive.

From this perspective our two cases, or more precisely their political contexts, differ substantially. Spain was governed by an authoritarian regime led by General Franco from the 1930s until the mid 1970s. As the Coto de Doñana National Park was created under that regime and strongly endorsed by the dictator himself, a lot of people in Andalusia still perceive the park as a heritage from a former era. Moreover, during these decades of dictatorship Spain was led in a centralist way. Andalusia, however, has a strong cultural and political regional tradition and, therefore, is characterized by a high degree of self-esteem. Under the dictatorship it could hardly deploy any of those characteristics. With democratization and subsequent devolution Andalusia regained some regional political identity. At the same time, though, these political changes have revitalized some traditional political divides between the old and the new regime, and between the centre and the region.

The historical and political sensitivities relating to the national park are not unimportant to an understanding of the Coto de Doñana case. And we can now understand better why the Coto de Doñana disaster led to such major controversies between national and regional authorities regarding their respective competencies and failures. It may also now be clearer that the green corridor was an important issue in the prestigious attempt of the regional government to establish its own nature conservation policy. The latter also explains the consensus forged at the regional level.

Our Dutch case provides quite a contrast. Here we find ourselves in a country with a consensual political tradition and style, which is also characterized by its well-elaborated environmental policy. The former refers to the long history of political tolerance and inclusiveness in Dutch politics. With regard to our case study it particularly refers to the fact that, from the early days of environmentalism in the early 1970s, the Dutch environmental movement has been acknowledged as a partner by the Ministry of the Environment and is even largely subsidized by it. The environmental organizations participate in a series of advisory bodies at national and local level (Leroy and van Tatenhove, 2002, pp.163–84)). This is a good example of an inclusive and cooperative political culture, which facilitates the addressing of environmental issues. Moreover, during the frequent interactions between government and the environmental movement, their representatives apparently developed a common language and share more or less similar definitions of the problems and political arguments. This may explain why the environmental movement at first went along with the arguments endorsing the Betuwe railway: traffic substitution, technological modernization and less pollution. A political system, we might conclude, can even be so inclusive that it incorporates its initial challengers.

In contrast to the Andalusian case, in which disagreements arose about who granted the permit and who should have controlled it (and failed to), Dutch environmental policy right from the start of the Betuwe railway debate had its procedures and responsibilities in place. Those procedures offered opportunities for the launching of appeals and other legal protests against the plan. Thus the well-elaborated character of Dutch environmental policy in a way provided leverage against the railway. The example illustrates how political openness may create its own criticism – without, however, substantially affecting the balance of power.

## 4.2  Differences in environmental concern

Societies and political systems differ with regard to the priority their members and elites give to a variety of societal and political issues such as employment, healthcare, migration, poverty, quality of life and, among many other things, the environment – that is, societies differ with regard to environmental concern. From our earlier discussion we can conclude that, very roughly, northern European countries reflect a higher level of environmental concern than their southern counterparts. These different levels of concern, it is often argued, relate firstly to some cultural traditions. This is illustrated by the traditional concern for landscape preservation in the UK, which is deeply rooted and institutionalized in a series of societal activities and political arrangements. Also, environmental concern relates to economic conditions and opportunities as, generally speaking, environmental concern particularly flourishes in relatively privileged economic situations. In addition, as suggested in the section above, environmental concern relates to political openness: traditionally dominant political cleavages, a particular electoral system and the political elite's attitude may considerably hinder or facilitate the expression of environmental concern.

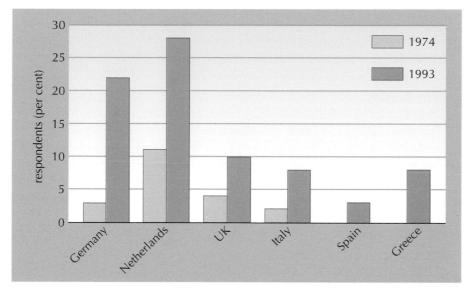

**Figure 4.9** Environmental concern in different EU member states: percentage of respondents who identified the 'environment' as the most important problem facing member states, 1974 and 1993.
*Source*: Weale et al., 2000, p.239.

We can see from Figure 4.9 that the countries in our two cases reflect some striking differences that help us to understand the dissimilarities in the cases. In Spain, environmental concern in general is clearly below the European average. Although there is no empirical evidence, for all kinds of reasons it is very likely that environmental concern is even lower in Andalusia. The weak position of the environmental movement in Andalusia is both a cause and a consequence of this below-average environmental concern. The Netherlands again displays a striking contrast as it belongs to the countries with the highest level of environmental concern, and the highest level of membership of environmental organizations.

Like political openness, however, environmental concern is a necessary but not a sufficient condition for effective societal action and political response.

## Summary

- Political power partly depends on the capacities of an actor to mobilize resources such as financial support, human resources, scientific expertise, and media attention. These resources facilitate an actor's capability to get things onto the public and political agenda.

- Political power, however, also depends upon characteristics that are beyond the reach of an actor, in particular upon the opportunities a political system offers, which refers firstly to the political openness of a system. Traditional political cleavages, the electoral system and the political elite's attitude are important features of this. The level of environmental concern is the second important factor providing political opportunities (or not). These political characteristics in a way mark the political space in which an actor is able to manoeuvre.

- The Betuwe railway case is an example of where environmental protest was able to mobilize and the context provided the opportunities to do so – and yet the protest was not that successful. In other cases those affected by environmental threats do not have the political power to act, because they do not have the capacity to sufficiently mobilize resources and/or because the political system does not provide the necessary political space in which to manoeuvre. The latter situation applied dramatically, for example, to the people living in the Chernobyl area.

- In those cases of relative powerlessness, the people affected depend upon political sponsors and supporters from abroad. This was seen in the Coto de Doñana case but in a negative way, as international organizations failed to put the issue onto the local agenda.

- As suggested by the reference to Chernobyl, some environmental issues and conflicts reflect this pattern of powerlessness in a far more distinctive way than our Spanish case, illustrating what is meant by environmental inequality. These cases vary from the local to the global: within cities one can often observe that those suffering from a poor environmental quality depend upon others for getting their claims onto the agenda. A similar pattern can be observed on a global scale, where poor countries and local protest groups need support from abroad in order to successfully fight their situation. These situations of an unequal environmental quality paralleled by an uneven political balance of power highlight the issue of environmental justice (see **Maples, 2003**).

# 5    The coming into being of environmental policies

So far we have focused on two cases, illustrating the process of agenda setting on specific issues in specific contexts. In doing so we related the agenda setting to broader and more general issues, particularly political power and its determining factors, environmental concern and related issues. It gave us an idea of the mechanisms by which a society responds to environmental threats and risks, and the factors that determine those mechanisms.

Building on the conclusions we have drawn from those cases we now turn to the way society and politics have responded to environmental issues as a whole. Earlier on we outlined the emergence of environmental concern from the early 1970s onwards. This section describes the early development of environmental policies, which we consider to be the gradual coming into being of a new field of political concern and a new domain of policy making. It led to the establishment of new governmental bodies and a series of legislative initiatives. The latter are used here to indicate a society's determination to handle environmental issues and to respond to them in a systematic way that becomes the subject of political debate. The content of the early environmental policy initiatives reflected the then dominant perceptions of environmental problems. Those perceptions – and the accompanying political responses and strategies – changed, as we shall see, in the 1980s and 1990s.

## 5.1 Ministries, laws and other institutions from the 1970s

Northern and western European countries established their environmental ministries mostly between 1970 and 1976 (see Table 4.2). That is, the gradual development of environmental policies occurred more or less simultaneously with the increasing environmental concern within both society and politics. Most southern European countries established their ministries about ten years later, from the early 1980s onwards (Weale et al., 2000). This delay does not only reflect the differences in environmental concern that we have seen earlier (Table 4.1 and Figure 4.9). When establishing their environmental policies, countries such as Spain, Portugal and Greece, primarily anticipated their membership of the EU. Literature on European environmental policy often differentiates between countries that are 'policy makers' – in particular Sweden, Denmark, the Netherlands and Germany – and others that are regarded as 'policy takers' – in particular Greece, Spain, Portugal and Ireland. Whereas the former states are all 'northern', the Irish example makes clear that the north–south divide should not be taken too literally. The UK does not fit into that pattern either, as it reportedly slowly evolved from being mainly a policy taker in the 1970s and 1980s to being a more active policy maker in recent years. Nonetheless, and without sticking to strict labelling or division, we can observe substantial differences in the speed and

**Table 4.2**   The establishment of environmental ministries

| | France | Germany | Netherlands | Sweden | UK | Spain |
|---|---|---|---|---|---|---|
| 1969 | | Interior Ministry acquires responsibility for environmental issues | | National Enviromental Protection Board Franchise Board for Environmental Protection | | |
| 1970 | | | | | Department of the Environment | |
| 1971 | Ministry for the Protection of Nature and the Environment | | Ministry of Public Health and Environmental Protection | | | |
| 1974 | | Federal Environment Office (research) | | | | |
| 1977 | | | | | | Ministry of Public Works and Transport becomes the main agency for the protection of the environment |
| 1986 | | Federal Ministry of the Environment | | | | |
| 1996 | | | | | | Ministry of the Environment |

*Source*: Aguilar Fernandez, 1998; Weale, 1992.

extent to which environmental policies are and have been developed in different countries.

From that same period onwards a series of environmental laws were developed – on air pollution, water pollution, waste management, soil sanitation, nature protection, noise pollution and other issues. While a succession of legislative initiatives might not seem very exciting, it should be realized that the first stages of environmental policy consisted primarily of such initiatives. The list of legislative measures in the UK, therefore, is quite similar to that from countries such as the USA, Canada, Sweden, Belgium, France, Germany, and the Netherlands. About ten years later, southern European countries followed that same track. Table 4.3 shows the legislative initiatives regarding air and water pollution and chemical waste in three countries. The list could easily have been enlarged with other environmental policy domains and with other countries. Despite the small sample, however, the list does show the late emergence of Spanish environmental legislation.

**Table 4.3**   Legislative initiatives in the early stage of environmental policy

|  | Netherlands | UK | Spain |
|---|---|---|---|
| Air | Air Pollution Act, 1970 | Health and Safety Act, 1974 | Law of Air Protection, 1972 |
| Surface waters | Surface Water Pollution Act, 1969 | Rivers Act, 1951; Control of Pollution Act, 1974 | Water Act, 1985 |
| Chemicals/ toxics | Chemical Wastes Act, 1976 | Deposit of Poisonous Wastes Act, 1972; Control of Pollution Act, 1974 | Law of Toxic Waste. 1986 |

*Source*: Aguilar Fernandez, 1998; Weale, 1992.

Simultaneous with the establishment of administrative bodies and the issuing of laws, most countries also set up agencies and institutes, such as monitoring institutes and scientific laboratories, for the underpinning of their environmental policies. This also led to the establishment of advisory bodies, such as the Royal Commission on Environmental Pollution in the UK. In some countries these scientific boards were complemented by consultative bodies to which representatives of market sectors (industry, agriculture) and societal organizations (environmental groups) were also invited. As we have seen above, the latter is the case in the Netherlands where environmental organizations, from the early 1970s onwards, have participated in a series of consultative bodies.

This phenomenon can easily be linked to our argument on political openness, as the participation of environmental organizations in these advisory bodies indicates a country's inclusive political attitude. It is worth noting, though, that the mere existence of advisory bodies regardless of who participates in them says nothing about the actual influence of these bodies and their members upon decision-making processes. But at least they provide political opportunities.

## 5.2  Early environmental policy: dominating conceptions of environmental issues

Lists of legislative measures and ministries indicate a society's determination to address the environment as a policy domain. They are part of the **institutionalization** of environmental policy. This refers to the fact that day-to-day behaviour is captured in patterns and that these patterns are, in turn, reproduced by day-to-day behaviour. In the case of environmental policy, insitutionalization refers to the ways in which policy is conceived, adopted and managed. Laws and governmental organizations reflect the hardware of institutions while problem definitions, strategic options and implementation issues constitute the software. As well as the institutionalization of organizations and administrative bodies, it is interesting to note the common pattern in the content of early political responses to environmental issues. They echo the then predominant conception of environmental problems and the predominant

insitutionalization

approach to tackling them. We will briefly indicate how these initial conceptions of problems have changed over time (van Tatenhove et al., 2000).

As we have seen from the lists of legislative initiatives in Table 4.3, environmental problems, firstly, were defined as problems of air pollution, water contamination, waste management and so on. In other words, they were approached sector by sector or medium by medium. Separate policies were developed on surface water, air pollution, household waste, industrial waste and other sectors. Also, standards were set sector by sector, and these were implemented by a system of permits. As legislation multiplied, we witnessed the coming into being of five, six or more permit systems, dealing with water, air and waste separately. It was not until the 1980s that the limitations and inconvenience of this so-called 'multiple permitting' approach were realized. This led to more comprehensive and more integrated environmental policies, based upon environmental quality standards that encompassed acidification, climate change and other major issues, no matter what sector of the environment was affected.

Second, these problems were mainly perceived as problems of emission overload that could be tackled by filtering and sanitation, in short, by 'end of pipe' technology. Consequently, the main strategy was the adding of filtering and sanitation technology, either to dissolve the pollution (for example, in air or water) or to gather and manage it on one particular spot (for example, water sanitation and waste management). Only later was it recognized that in most cases dissolution simply meant diversion rather than solution of the problem. Acidification, again, serves as an example: building higher smokestacks actually meant the export of emissions, for example from British industries into Scandinavian lakes or from Czech and Polish industries into German woods, causing the so-called *Waldsterben* (the devastation of woods). As a consequence, environmental policies evolved, with the aim of controlling pollution at source: instead of building higher chimneys, they aimed at decreasing the volume of emissions and, eventually and when necessary, at decreasing the volume of production itself.

The policy on pollution caused by cars also illustrates this shift, since it evolved from adding catalytic converters to aiming at reducing the number of car-kilometres. However, although the latter strategy seems far more effective and therefore preferable, politically and socially it represents the hardest strategy, since it tends to affect directly people's preferences and behaviour. Decreasing emission volumes and filtering them in the long run might be environmentally less effective but it is, as it was in the early 1970s, politically the easier strategy.

Third, early environmental policy initiatives mainly if not exclusively dealt with local and regional issues, that is, with 'hot spots' – particularly polluted areas threatening people's health or relatively quiet areas that came under pressure from industry or urbanization. From the mid 1980s onwards this predominant conception of the problem rapidly changed as acid rain in particular illustrated the cross-boundary effects of environmental threats and showed the need for transnational policies. We have seen a geographical widening in the scope of environmental policies since then, ending up with the agenda setting of global issues such as climate change and biodiversity from the 1990s onwards discussed in Chapters Five and Six.

### Activity 4.7

In this section we identified three main features that characterized the early stages of environmental policy. Can you explain these three, and elaborate on how they gradually became regarded as shortcomings?

### Comment

- Environmental problems were approached sector by sector or medium by medium. This leads to a fragmented policy since individual activities need to acquire a series of permits for their respective forms of pollution. This practice creates an administrative labyrinth and huge inefficiencies, particularly since the responsibilities are divided among a number of administrative bodies. On top of this the administrative system becomes overwhelmed by the thousands of permits it has to deal with.
- Environmental problems were perceived essentially in terms of emission overload that could be solved by adding 'end of pipe' technology. When some more complex problems such as acidification became manifest, clearly illustrating the diffuse, cross-sector and transnational character of the issue, the failure of the approach became clear.
- Environmental policy initiatives mainly dealt with local and regional issues. When acidification, desertification, the depletion of the ozone layer, the climate changes and other changes became prominent issues, it was clear that some environmental problems occur on a larger geographical scale. These problems call for the widening of the geographical scope, and particularly for an international approach.

## Summary

This brief overview of the early political response to environmental problems leads to two main conclusions:

- Whereas environmental policy initiatives in north-western European countries date from the early 1970s, most southern European countries initiated their environmental policies in the 1980s. The EU seems to have played a key role, having accelerated environmental policy initiatives in the latter countries.
- Notwithstanding some national differences, the initial political responses to the environmental challenges echo quite similar conceptions of environmental issues and the best way to tackle them. The initial stages of environmental policy essentially consisted of a series of state-initiated

regulatory measures, dealing with environmental issues medium by medium, and handling them primarily as 'end of pipe' problems. Simultaneously, the implementation and monitoring of these policy measures led to the establishment of a series of governmental bodies.

# 6   The renewal of environmental policies

Although we cannot make clear distinctions about ongoing historical processes, we can argue that the first stage of the institutionalization of environmental policies lasted about a decade and a half, in other words from the early 1970s to the mid 1980s. From that time a gradual but substantial renewal of environmental policies took place (Weale, 1992). In this section we explore the whys and hows of that renewal.

## 6.1   Some deficiencies of early environmental policy strategies

From the mid 1980s onwards, the shortcomings of the then dominant environmental policies gradually became obvious. First, some management failures arose. In particular the 'multiple permitting' allowed individual industries to acquire a permit for several forms of pollution, the regulatory responsibilities for which were divided among a number of administrative bodies. As noted earlier this created an administrative labyrinth, huge inefficiencies and a tremendous overload of the state bureaucratic system's capacity.

From our Coto de Doñana case we learned that neither national nor regional authorities succeeded in forcing the Boliden company to comply with the standards set out in an environmental permit. In the early 1980s there was scientific evidence that even in countries where environmental policies are well developed, such as Denmark, Germany and the Netherlands, only about 25 per cent of the activities that needed a permit actually had one, at least one that was appropriate. The permit system was obviously not a perfect strategy for the management of environmental issues, and therefore complementary and alternative policy strategies were sought.

Second, the then dominant approach, in which air, water and other media were treated separately, was questioned as it turned out to be inappropriate for tackling cross-media transfers, or for inducing the necessary changes within the production processes themselves. As mentioned earlier, 'acidification' is a good example of the issues at stake here. Acidification does not refer to a specific polluted medium, but to processes that affect water, air and soil simultaneously and thereby threatens ecosystems as a whole. Moreover, acidification is caused by different pollutants such as sulphur dioxide ($SO_2$), ammonia ($NH_3$) and nitrogen oxides (NO$_x$) and these pollutants are produced respectively by industry, agriculture and traffic. The response to acidification thus illustrates the inadequacy of a political response that applies thousands of permits to individual activities and to those pollutants separately, while three main sectors of contemporary society are clearly responsible. Acidification, it gradually became clear, required an overall policy aiming to reduce the amount of sulphur in energy production and industry in general, to reduce the amount of nutrients in the agricultural sector, especially in intensive cattle agriculture, and to decrease car traffic. In other words, instead of fighting emission levels, one needed policies aimed at changing industrial and agricultural production processes and decreasing the fluxes of possible pollutants they create.

## 6.2  Regulatory instruments complemented

As these shortcomings were discussed, albeit in nationally different terms, they led to quite similar results. One answer, clearly, was better management, in other words the co-ordination of fragmented permits and the integration of environmental policies, in most countries leading to 'green planning'. These plans introduce long-term and short-term policy goals in order to overcome the proliferation of goals and standards. They all try to integrate sector-specific policies by introducing general objectives such as 'sustainability'.

But the process went further. As we have seen, environmental policies initially consisted mainly of state-initiated regulatory measures. With this so-called 'direct regulation', governments aim at imposing regulations on citizens and companies by establishing legislation, thereby hoping to influence directly citizens' and companies' behaviour. As the limitations and shortcomings of this approach became obvious, a diversification of policy strategies was advocated aiming at complementing regulatory instruments with economic ones. Economic instruments essentially connect financial incentives and penalties to environmentally friendly and unfriendly behaviour respectively. Either they reduce the cost of the desired activity or they increase the competitors' costs, or they do both. Levies and subsidies directly encourage or discourage the environmentally relevant behaviour one wants to influence. Taxes are particularly used to increase the price of environmentally harmful products. These economic instruments are labelled 'indirect regulations', as they do not build on coercion, but on the consumers' and producers' well considered self-interest – mere economic calculation is supposed to make them choose the environmentally friendly alternative. In other words, government uses the carrot instead of the stick to influence societal behaviour.

Economic instruments are discussed in Chapter Three.

The use of communicative instruments or 'social regulation' builds not on an actor's well-considered self-interest, but primarily aims at appealing to an actor's social responsibility. The supply of information about the issue, about somebody's share of responsibility and about their possible contribution to resolve it, should provoke compliance with policy goals. As far as citizens are concerned the underlying assumption is that the lack of adequate information about both their share in the problem's cause and in its solution hinders the necessary behavioural change. Therefore information should be the conveyor of behavioural change. Eco-labelling of different sorts (see Figure 4.10) is one of the well known examples of this.

Table 4.4 summarizes the advantages and disadvantages of the three kinds of policy instruments. Each of the approaches in Table 4.4 is based on certain assumptions about behaviour which may not always occur in practice.

**Table 4.4**    The advantages and disadvantages of different sorts of environmental policy instruments

|  | Legislative instrument | Economic instrument | Communicative instrument |
| --- | --- | --- | --- |
| **Advantages** | Clear and univocal rules and standards | Appealing to consumers' and producers' well considered self-interest | Appeals to people's own conviction and responsibility |
| **Disadvantages** | Causes a lot of administration and bureaucracy (permits, control) Presumes enforcement and coercion by state representatives | Presumes rational, even calculable choices between behavioural alternatives High and hardly predictable costs (in the case of subsidies) | Presumes that information and communication produce effect |

Direct regulation presumes enforcement and coercion by state representatives. But it depends on the state's capacities to manage and control some thousands of permits, which clearly illustrates the limits of its direct regulating capacity.

Economic regulation is discussed in Chapter Three.

Economic regulation presumes rational and calculable choices between behavioural alternatives. But do you really calculate the price of a car ride, and do you compare that systematically to the cost of a bike ride or a walk? In many cases these costs are hardly calculable, and even when they are, you could prefer a certain traffic mode for non-cost reasons, for example prestige, comfort and relaxation.

Communicative regulation presumes that information and communication produce effects on behaviour. Smoking is a good counter-example to illustrate the limits of this assumption: there is hardly any issue that has had so much attention in public communication campaigns over the last 20 or 30 years, and yet the number of smokers has not decreased accordingly. The influence of information is, among other things, limited by mechanisms such as selective perception and non-cognitive considerations (tradition, prestige, social pressure and so on). Similar mechanisms also apply to our traffic behaviour and to the way we deal with waste.

**Figure 4.10** The EU Eco-label.

## 6.3 Changes in the state–market–civil society interrelations

Although quite important, the 'instruments' debate tends to reduce environmental policy simply to a management issue, as if society's capacity to respond to environmental threats merely depends upon the effectiveness of its policy instrumentation. Beyond the instrumentation issue, however, is a politically more interesting question: who is primarily responsible for the necessary societal changes and for the organization of a proper political response? In the early days of environmental policy it seemed as if the state, quite naturally and without too much questioning, had to take the lead in the design and implementation of political responses. From this perspective, the instrumentation debate only addressed the question of which strategy the state could use to mobilize power and legitimacy to initiate the necessary changes. But, given the all-encompassing changes in our society, this increasingly seems insufficient. The changes we refer to go far beyond environmental politics and policies, but we will restrict ourselves to those changes that immediately affect these policies.

From the early 1980s onwards a mixture of facts and ideologies provoked a fundamental political debate on the role of the state throughout the democratic world. The 'facts' were that, not only in environmental policies, the idea of a manageable and governable society for which governments could create blueprints and steer future developments was gradually eclipsed. Governmental initiatives in a variety of policy fields fell short and, therefore, could not keep up with the optimism of the 1960s and 1970s regarding the steering capacity of the state. The 'ideologies' were advocated by some neo-liberal politicians who made their way to the centres of political power: President Reagan in the USA and Prime Minister Thatcher in the UK were the most prominent of them. They claimed that the state's responsibility and role should be limited, giving (back) more autonomy and responsibility to both the market and **civil society**, that is, civil society citizens and their social and cultural organizations.

Within this context neo-liberal programmes have been realized, for instance in literal retreats of the nation state by the privatization of former state-owned public facilities, such as communications and public transport. Within the environmental field, these policies are reflected in the privatization of drinking water supply, waste-water management, waste management and others. France, the UK, Belgium and the Netherlands all privatized some of these public facilities. The arguments for this privatization were manifold, but essentially came down to cuts in public spending (Belgian waste-water management), to restricting the role of the state to its core business (Dutch waste management), and anticipating a better performance from private companies in the fields mentioned (UK's drinking water supply). Similar arguments were brought forward for initiatives on deregulation and devolution to autonomous agencies that often accompanied privatization efforts.

Privatization, deregulation and devolution were among the most spectacular political changes at national level in the 1990s. They clearly implied a step back of the state in favour of private market agencies and they thereby replaced considerations of access to public facilities with market mechanisms. However, while water supply was privatized it was not deregulated and in England and Wales there are no less than three regulatory agencies, namely, the Drinking Water Inspectorate, OFWAT (cost/price regulator) and the Environment Agency. This mix of private companies and public agencies indicates a shift in the balance of power which does not facilitate citizens' influence. This combination of privatization and regulation by agencies which has occurred in several sectors, notably public transport, the generation and distribution of energy, the removal and processing of waste and others has put them largely beyond the influence of both politicians and citizens, while consumer control is hardly in place.

Privatization clearly affected the organization of environmental policies as well. Quite obviously the state's power over privatized or 'autonomous' agencies was less than before. Apart from the state's loss of direct control, we have also witnessed a decrease in the state's power in the setting of standards. Instead of setting standards based upon the advice of scientific boards and trying to implement those standards via permits, governments nowadays increasingly turn to the negotiating table with specific branches of industry (Carraro and Lévêque, 1999; Glasbergen, 1998; Mol et al., 2000). In these talks representatives of both parties aim at setting goals that form a compromise between what is ecologically needed and what is economically achievable. These negotiations often lead to covenants or gentlemen's agreements between governmental bodies and market representatives on the when and how, and on who is responsible for initiating, implementing and monitoring a systematic change in production processes and product qualities. Therefore, horizontal mechanisms replaced hierarchical mechanisms, negotiated standard setting replaced unilateral standard setting, and certification and self-regulation replaced (failing) state inspection. This led to new instruments and standards, largely initiated by industry itself, at national, European or global level.

This change in state–market interrelations, some argue, has been paralleled by a similar change between state bodies and civil society. Here again governmental bodies find themselves faced with the non-acceptance and non-implementation of policies. Particularly in controversial issues, such as the siting of energy plants, the extension of airports, and major infrastructural works, governments face societal mobilization and protest, leading to legal procedures. This often results, as we have seen in the Betuwe case, in huge delays or cost increases, or in both. Here again governments have realized that state power does not suffice for the realization of such plans, as state power in the end is conditioned by legitimacy.

# 6.4 Consensus building and its limitations

To overcome such deadlocks, state representatives nowadays invite the stakeholders involved, including market agencies and civil society representatives to the negotiating table, aiming at building a consensus on possible solution strategies and decisions. Consequently, all over Europe these days we witness different variants of this approach, labelled 'participatory', 'interactive', or 'deliberative' and so on. Whatever their names, they all aim at consensus among all parties involved, and they are conceived as methods of conflict resolution. Particularly with regard to the extension of airports, from London Heathrow to Vienna, and from Amsterdam Schiphol to Paris Charles de Gaulle, we can witness such approaches, including the deployment of an impressive arsenal of mediation techniques. The way the regional government of Andalusia has set up the decision-making process for the green corridor followed quite a similar 'participatory' track, with representatives from governmental bodies, market agencies (agriculture, tourism) and civil society (local citizens) around the table.

Deliberative approaches are discussed in Chapter One and in **Burgess (2003)**.

While the latter was clearly innovative in southern Spain, these approaches are most frequently used in countries with a tradition of political consensus, such as the Netherlands and Austria. Hence the lack of participation in the decision making on the Betuwe railway is surprising, and caused huge protests and costs as we have seen. That is the more surprising since in other cases, including the extension of both the Rotterdam harbour and of Schiphol airport, a much more participatory approach was put into place. But it might not be surprising when we interpret the Betuwe railway itself as predominantly a private market initiative endorsed by the state, that is, it is an example of changed state–market power relations. As the era of massive state budget cuts seemed to have come to an end in the late 1980s, we witnessed a series of huge investments, particularly in infrastructure all over Europe: airports extensions, high speed trains, tunnels and bridges. Due to privatization and the crucial impetus to the economy that these investments were claimed to represent their construction could not afford to be delayed. Public participation was believed to cause too much delay, which is why we see old-fashioned, non-participatory decision making dominate, exactly where market imperatives prevail. Therefore, and despite the efforts on participatory decision making, the Betuwe railway case illustrates how the change in the balance of power between state and market, seems to decrease the influence of civil society. This brings us back to one of the paradoxical conclusions of the Betuwe railway case mentioned above: despite all legal and institutional opportunities for participation that were created in the Netherlands and despite the fact that the protesters in this case have used almost all of these, their political power over the actual decision making was very limited.

## Summary

In this section we discussed two, at first sight, paradoxical trends in environmental policies of the 1980s and 1990s, which led to questions about political power, particularly about the balance of power between state, market and civil society.

- Aiming at a more effective strategy, governments first looked for a more diversified strategy and complemented their initial regulatory strategies with economic and communicative instruments. More or less simultaneously and due to a more encompassing political debate on the state's role, the latter was actually restricted, primarily in favour of market agencies.

- This change in the organization of, among others, environmental policy results from the increasing awareness that, put negatively, the steering capacities of governmental bodies are quite limited. Put positively, it results from the awareness that governmental bodies and market agencies should share their responsibilities to design and implement the necessary changes, and that an effective policy will only result from their common efforts.

- The evolution towards a less hierarchical and more negotiated environmental policy leaves us with some issues that go beyond the question of whether this policy is more or less effective. First, just as with the public, industry's environmental awareness and willingness is also not as widely spread as is sometimes suggested. Firms and their subsidiaries often behave quite differently in different political contexts, as we know from Shell's and other western companies' behaviour outside Europe. Where societal and political pressures are lacking, industry often fails to comply with environmental standards. The Coto de Doñana case illustrates the issue, and there is overwhelming evidence of even worse cases in the developing world. Second, the newly deployed arsenal of negotiated environmental policy making may have enlarged people's opportunities to participate, yet it does not increase people's political power in environmental policy making, as the Betuwe railway case made clear.

# 7    Conclusion

The main issue in this chapter was the question of how to bring about the societal changes that are necessary to tackle successfully environmental problems – in other words, how to mobilize and organize society's capacity to respond to these problems.

Starting from two cases we acquired some insights into the processes of agenda setting. We learned that these processes depend upon the actors' capacities to mobilize resources, as well as upon societal, economic and political circumstances. We could list some of these political factors and features that constrain or enable opportunities for protest and affect its impact, that is, we could list some of the factors that determine political power balances. We stressed the political opportunities that a certain political system provides, or does not, by its political openness, the way it organizes public participation in search of public support, and political legitimacy. We have also seen that environmental concern is an important feature, as it initiates or enables social mobilization as well as the organization of administrative capacity for legislation, monitoring and other basic equipment for an environmental policy to come into being.

We then learned about the early stages of environmental policies, the initial strategies and the conceptions of the issue behind them. We emphasized the changes in instruments and organizations that characterized the recent evolution of environmental policies. These changes should be regarded as a result of a more comprehensive political change in western societies, and should be assessed in terms of power balances and their possible and actual shifts. The latter led on to the conclusion that these recent political changes, particularly the shifts in the balance of power in favour of the market, do not seem to be favourable for the empowerment of environmental concern, for societal mobilization and political response.

We do not want, however, to conclude this chapter on a pessimistic note regarding society's power and capacity to respond to environmental issues. Compared to the early 1970s, we are witnessing a still growing level of environmental concern. Yet environmental concern alone is not sufficient to produce a political response. It is true that environmental issues increasingly have become part not only of state policies but also of private firms' marketing strategies, where they provide competitive advantages in some cases. It is also true that we have witnessed an increase in valuing and pricing the environment, for example by taxes on the use of natural resources, waste, and traffic. Yet it is hard to see how mere market mechanisms as such could ever remedy all the environmental issues we face.

The same applies to a solely state-controlled environmental policy since, as we have seen, the societal changes needed far exceed the state's steering capacities. The latter seems to be reinforced by another major trend that we have not discussed here, although it parallels privatization, and that is globalization. It would not be possible in a few lines, however, to explain thoroughly what globalization is and what its possible and actual impact on the environment could be. We restrict ourselves to suggesting that globalization, regarded here as essentially the global spreading of economic activities looking for the most profitable location, is almost entirely beyond political control, let alone societal control. Even as environmental concern is spreading among international organizations such as the IMF, the World Bank and the WTO, these organizations actively endorse economic globalization, including those subsequent actions and decisions that cause major environmental harm. And these and other international organizations have an increasing influence upon the environmental policy of nation states, whether members or not. These organizations do not face too much societal participation and control, as the civil society's political participation is still mainly organized at local and national levels.

This is exactly why internationally operating NGOs such as the WWF, Greenpeace and others, in concert with the recently emerged anti-globalization movement, campaign for the establishment of democratic rules and of opportunities for participation at global level. The more our economy and our politics globalize, the more need there is for society's participation at that level, to ensure a less uneven balance of power (see Chapter Five and **Taylor, 2003**).

Compared to the cases set out in this chapter, at international level we are witnessing only the first steps in the process of agenda setting of environmental issues, and have yet to see the emergence of systematic environmental policies, endorsed and empowered by societal mobilization and leading to firm political responses. From our cases we learned how difficult the empowerment of societal action is, and what circumstances are needed to bring about a successful political response. In our cases in the Netherlands and Spain or in Europe in general, even where there is some environmental concern, where scientific and administrative capacity are in place, and where there is some constitutional guarantee for civil society's political influence, there is still a long way to go.

# References

Aguilar Fernandez, S. (1998) 'New environmental policy instruments in Spain' in Golub, J.S. (ed.), *New Instruments for Environmental Policy in the EU*, London, Routledge.

Bachrach, P. and Baratz, M. (1970) *Power and Poverty*, Oxford, Oxford University Press.

**Bingham, N., Blowers, A.T. and Belshaw, C.D. (eds.) (2003) *Contested Environments*, Chichester, John Wiley & Sons/The Open University (Book 3 in this series).**

Blowers, A.T. and Leroy, P. (1994) 'Power, politics and environmental inequality: a theoretical and empirical analysis of the concept of "peripheralisation"', *Environmental Politics*, vol.3, no.2, pp.197–228.

**Blowers, A.T. and Elliott, D.A. (2003) 'Power in the land: conflicts over energy and the environment' in Bingham, N. et al. (eds).**

**Burgess, J. (2003) 'Environmental values in environmental decision making' in Bingham, N. et al. (eds).**

Carraro, C. and Lévêque, F. (eds) (1999) *Voluntary Approaches in Environmental Policy*, Dordrecht, Kluwer Academic Publishers.

Carson, R. (1991) *Silent Spring,* Harmondsworth, Penguin (first published in 1962).

Dutch Bird Protection Agency (1999) *Vogels*, November/December, no.6.

**Freeland, J.R. (2003) 'Are too many species going extinct? Environmental change in time and space' in Hinchliffe, S.J. et al. (eds).**

Glasbergen, P. (ed.) (1998) *Co-operative Environmental Governance – Public–Private Agreements as a Policy Strategy*, Dordrecht, Kluwer Academic Publishers.

Hannigan, J. (1995) *Environmental Sociology*, London, Routledge.

**Hinchliffe, S.J., Blowers, A.T. and Freeland, J.R, (eds) (2003)** *Understanding Environmental Issues,* **Chichester, John Wiley & Sons/The Open Unviersity (Book 1 in this series).**

Jongman, R.H.G. (1995) *Landscape Ecology and Land Use Planning. The Use of Landscape Ecology in Land Use Analysis and Planning for Nature in Europe*, Landbouwuniversiteit Wageningen.

Kingdon, J.W. (1984) *Agendas, Alternatives and Public Policies*, Boston/Toronto, Little, Brown and Company.

Kriesi, H.P., Koopmans, R., Duyvendak, J.W. and Giugni, M.G. (1995) *New Social Movements in Western Europe – A Comparative Analysis*, Minneapolis, University of Minneapolis Press.

Leroy, P. and Tatenhove, J. van (2002) 'The environment and participation' in Driessen, P. and Glasbergen, P. (eds), *Greening Society*, Dordrecht, Kluwer Academic Publishers.

**Maples, W.E. (2003) 'Environmental justice and the environmental justice movement' in Bingham, N. et al. (eds).**

Meadows, D.H., Meadows, D.L., Randers, J. and Behrens, W.W. (1972) *The Limits to Growth*, London, Earth Island.

Mol, A.P.J., Lauber, V. and Liefferink, D. (eds) (2000) *The Voluntary Approach to Environmental Policy*, Oxford, Oxford University Press.

Pestman (2001) *The Betuweroute: Mobilization, Decision Making and Institutionalisation on a Large Infrastructural Project*, Nijwergen, Amsterdam, Rozenberg Publishers.

**Smith, J.H. (2003) 'Making environment news' in Bingham, N. et al. (eds).**

Sunyer, C. and Manteiga, L. (1998) *Financial Instruments for the Natura 2000 Network*, TERRA Environmental Policy Centre.

Tatenhove, J. van, Arts, B. and Leroy, P. (eds) (2000) *Political Modernisation and the Environment. The Renewal of Environmental Policy Arrangements*', Dordrecht, Kluwer Academic Publishers.

**Taylor, A. (2003) 'Trading with the environment' in Bingham, N. et al. (eds).**

Weale, A. (1992) *The New Politics of Pollution*, Manchester, Manchester University Press.

Weale, A., Pridham, G., Cini, M., Konstadakopulos, D., Porter, M. and Flynn, B. (2000) *Environmental Governance in Europe – An Ever Closer Union?*, Oxford, Oxford University Press.

# Answer to Activity

(a) Alarming publications, environmental disasters and specific images.

(b) Alarming publications: *Silent Spring*, published by Rachel Carson in 1962, sounded the alarm over the accumulation of insecticides, DDT in particular, in the food chain; *Limits to Growth*, published by the Club of Rome in 1972, warned about the rapid depletion of natural resources.

Environmental disasters: the sinking of the Torrey Canyon in the Channel in 1966, among many others from that period.

Specific photographs: the US Apollo space programme provided photos of Earth seen from the moon, in which our planet looked beautiful but very fragile.

(c) The publications provided 'scientific evidence' on the seriousness of the environmental issue, and were therefore quite convincing, since they forecast serious problems if society failed to come up with an adequate response. The accidents in a way reinforced the same message, as they seemed to prove the dramatic expectations. The pictures of the unique and fragile Earth became very sensitizing, since they made visible what was beyond sight till then: the beauty of the planet, the physical evidence of the 'global village', and its vulnerability.

# Chapter Five

# Climate change: global responses under uncertainty

Stephen Peake

# Contents

# 1    Introduction

The world's nations are far from 'united' on how best to respond to climate change. If they succeed in averting the worst climate predictions it will be an unprecedented feat of human achievement: environmental management on a global scale. The consequences of failure could be very severe for human society. Such failure would mark the limits of our capacity to cooperate, organize and act at the international level to manage environmental inequalities and impacts.

It is hard to imagine a better example than climate change for thinking through the complex and dynamic forces that shape environmental change at the global level. These forces are physical, ecological, social, economic, technological and political. Predicted climate change is in large measure a response to human activity, and there are uncertainties and risks if we do not take **precautionary action**.

precautionary action

The precautionary principle is defined in **Blowers and Smith (2003)** and discussed in **Bingham and Blackmore (2003)**.

There are also, however, risks if we do. The risks are only in part 'environmental' – they are also economic and social. Choosing the best course of action to respond to climate change is forcing policy makers to confront varying degrees and types of uncertainty.

There are many different perspectives on the nature of risks involved with climate change. Countries find themselves bound up in climate politics: some worrying about how much land they will have left by the end of the century; some fearing permanent drought and famine; others, who are fossil fuel dependent, wondering what will happen to their economies and lifestyles if renewable technologies take over and the global oil market collapses; and some simply pondering the economic consequences of a low-carbon, low-consumption 'sustainable' future (Figure 5.1).

The nature of the risks we face are unevenly spread in both space and time and are highly uncertain: our current lifestyles and aspirations have become inextricably linked to distant others and tangled up with future events. There are three good reasons for this pervasive uncertainty:

Chapter One discusses uncertainty.

- There are gaps in our understanding of how the climate system functions.
- Uncertainties are inherent in forecasting changes in a complex and non-linear system such as the Earth's climate.
- Great uncertainty surrounds the nature of our own human reaction to climate change in the coming years.

Human intervention is potentially a powerful negative feedback to influence the climate system. 'Negative' feedback is a very 'positive' factor in restoring and maintaining balance in both living and non-living systems (you came across feedbacks to maintain temperature control in heating systems in Chapter Two). It could limit damage and reduce the rate of climate change. That is, however, providing we can gauge the strength and timing of our responses in the face of considerable scientific uncertainty and potentially unknowable risks.

**Figure 5.1** The impacts of climate change and the cost of implementing policies to act on it are felt differently across the world: a tropical monsoon, Thailand; a trafic jam in London; and an oil installation in a Gulf State.

In this chapter, I aim to use the examples of the United Nations Framework Convention on Climate Change (UNFCCC) and its Kyoto Protocol to show the interplay at the international level of the theme of responses and the concepts of risk and uncertainty (scientific and political). These are good examples of the tools and approaches that are available for managing global environmental problems. Climate change is also a very practical example of the constraints and opportunities for achieving the broader goal of sustainable development. Deciding how best to respond to climate change is about thinking through environmental change under extraordinary conditions of scientific and political uncertainty. The Intergovernmental Panel on Climate Change (IPCC) has characterized climate change decision making as essentially 'a sequential process under uncertainty'.

The chapter begins by considering some basic but important policy-relevant questions relating to the nature of the scientific and political uncertainties and risks we face. We go on to consider three basic options for responding to climate change (in section 3) and the nature of the uncertainties these options, as feedbacks, invoke (in section 4). In section 5 you are introduced to the UNFCCC and its Kyoto Protocol as examples of international legal instruments within the rapidly developing domain of global environmental governance.

I hope that by the end of this chapter you will have a deeper understanding of decision making under uncertainty and why it matters. My aim is to stimulate you to think critically about how environmental responses at the international level work in both theory and practice.

# 2    A climate of pervasive uncertainty and risk

Our scientific understanding of climate change is improving rapidly, but it is far from complete and uncertainty is pervasive. Thousands of scientists continue to unravel the complexities of the many couplings (for example, the behaviour between the atmosphere and the oceans) and feedbacks (for example, the positive feedback of water vapour) among the biogeochemical subsystems that determine the overall behaviour of the climate (**Blackmore and Barratt, 2003**). Many scientists are busy gathering new data and building complex computer models of the climate system.

From a policy-making perspective, deciding the best course of action to respond to the threat posed by climate change raises basic questions on the causes, effects and consequences of climate change. The most comprehensive and influential source of this evidence is synthesized in the work of the Intergovernmental Panel on Climate Change (IPCC). The IPCC's 2001 Third Assessment Report (IPCC TAR) is our best available source to answer policy-relevant questions such as:

- Is the climate changing and how significant is this?
- Why is the climate changing?
- What is the scale of future climate change?
- What are the risks, consequences and costs associated with future climate change?

The science flowing through the answers remains highly uncertain and contested. Uncertainty is compounded from one answer to the next and cascades towards perhaps the most important question of all: what is the nature of the risk we are facing? Decision making rests on two broad perspectives – take action to slow down the rate of climate change or 'wait and see'.

For some observers, the scientific claims about future climate risks are already enough and beg urgent and drastic precautionary measures in order to restore thermal equilibrium to the planet. These measures include reducing emissions of greenhouse gases and taking actions such as building sea walls to adapt to anticipated climate change. For others, the contested aspects of much of the science provide a reason to wait and see, to respond slowly by adapting to or coping with climate impacts as they unfold. These two perspectives are at the heart of the political tensions and uncertainties among nations and other stakeholders. Both scientific and political uncertainties are involved in climate change, and they are connected in various ways. Box 5.1 identifies ten characteristics of the climate change issue.

## Box 5.1  Climate science and politics: ten sources of uncertainty and risk

- The problem is global. No single individual or nation can determine the composition of the world's atmosphere.

- There are multiple players (governments, multinational firms, local enterprises, citizens) and therefore multiple challenges for, and limits to, collective actions and responses.

- The scope of human activities that cause climate change is all-encompassing (from, say, wood fuel gathering in rural African villages to life in one of the world's megacities).

- The problem is long term. It is greenhouse gas *concentrations* that matter and not the amount of emissions at any one time. The inertia of the climate system means that decisions taken by the present generation will affect future generations.

- Responsibility for historic contributions to the problem of climate change is unevenly spread. The climate change we are now observing is the result of the Industrial Revolution in the West. However, the climate of the twenty-first and twenty-second centuries will be dramatically affected by developing countries today, including some of the world's most populous countries such as China and India. This is an important force for *political mobilization* on the issue (see Chapter Four).

- The impacts of climate change are unevenly spread. Some of the poorest countries of the world are extremely sensitive and vulnerable to climate change. This is another important force for political mobilization.

- There is no single technological panacea or policy response that could slow down the rate of climate change.

- Responses will involve many technologies, policies and measures and therefore complex interactions between them. The management of the intentional and unintentional consequences of responding presents a considerable challenge for environmental management at the global scale.

- Climate change involves potentially irreversible consequences.

- The global institutions necessary to respond are young, or not yet born.

(adapted from IPCC, 2001c)

To begin to understand the political forces gathering around the issue of climate change, we must first understand how political uncertainty is partly rooted in scientific uncertainty. How well can we answer the policy-relevant questions stated above on the basis of the best available scientific evidence provided by IPCC TAR?

## 2.1 Is the climate changing and how significant is this?

Global mean surface temperature (GMST) is the most widely used proxy indicator of climate change. There is clear paleoclimatic evidence of warm interglacial and cold glacial periods about every 100,000 years, with global temperature swings of up to 8 degrees. We also know that global temperature has continued to vary by up to 2 degrees in the last 10,000 years (a relatively stable and warm interglacial period). Over the last 140 years, direct thermometer measurements indicate warming of 0.6 ±0.2 °C (Figure 5.2) (**Blackmore and Barratt, 2003**).

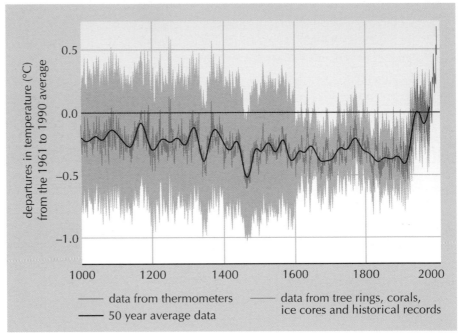

**Figure 5.2** Scientific evidence of the variation of the mean surface temperature in the northern hemisphere over the past 1,000 years. (Variation over the whole globe cannot be seen as there is insufficient proxy evidence for the southern hemisphere.) The grey area represents a 95 per cent confidence range in the annual data.
*Source*: adapted from IPCC, 2001a, Figure 2.1.

So, yes, we are quite certain the climate is changing. Indeed, it would be headline news if this were not the case. According to the IPCC (IPCC, 2001a), 'Globally, it is very likely that the 1990s was the warmest decade and 1988 the warmest year, in the instrumental record (1861–2000)'. There are some remaining disputes about recent temperature observations, including concerns that measurements are distorted by urban 'heat island' effects, that the network of surface monitoring sites is not truly global and that the data conflict with recent satellite observations which in fact show a cooling trend. Weather indicators do however suggest a warming. The IPCC notes that 'changes in sea level, ice extent and precipitation are consistent with a warming climate near the Earth's surface'. Indicators of changes in physical and biological systems also point in the same direction

**Figure 5.3** Biological systems and some extreme weather patterns are changing in directions compatible with a warming world: a starving polar bear (*Ursus maritimus*) in fish-depleted Arctic waters; homes threatened by fire on a Cape Town beach; and frogs (*Rana Temporaria*), a UK species, spawning two weeks early.

(Figure 5.3) and 'observed changes in regional climate have affected many physical and biological systems, and there are preliminary indications that social and economic systems have been affected'. This leads to the IPCC's main conclusion that 'an increasing body of observations gives a collective picture of a warming world and other changes in the climate system'.

## 2.2 Why is the climate changing?

Our confidence in answering this question is growing, though considerable uncertainty remains. A half degree warming during the Industrial Revolution is not particularly significant relative to variations over the last 1,000 years. Even so, according to the IPCC, 'most of the warming observed in the last 50 years is attributable to human activities'. The smoking gun is the correlation with the rapid and significant build up of greenhouse gases in the atmosphere as a result of fossil fuel combustion and land use change (Figure 5.4).

We have direct measurements of greenhouse gas concentrations since the 1950s and proxy indicators going back a thousand years. There are few, if any, disputes about observations of rising greenhouse gas concentrations. The various flows in the carbon cycle (for example, cumulative emissions from fossil fuel combustion into the atmosphere, and the absorption of carbon dioxide from the atmosphere by oceans and forests) are reasonably accounted for (to within 10–15 per cent) and levels of uncertainty are moderate; the IPCC concludes that human activities

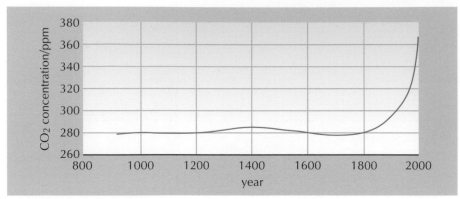

**Figure 5.4** The smoking gun: graphic scientific evidence of the increase in atmospheric carbon dioxide concentration over the last 1,200 years. The concentration remained stable until about 1750, when it started to rise exponentially.
*Source:* adapted from IPCC, 2001a, Figure 3.2.

An aerosol is a collection of tiny liquid or solid particles dispersed in a gas; e.g. dust in the atmosphere.

See **Blackmore and Barratt (2003)** for a discussion of the enhanced greenhouse gas effect and radiative forcing.

have increased the atmospheric concentrations of greenhouse gases and aerosols since the pre-industrial era. The enhanced greenhouse effect provides a credible explanation of the link between rising greenhouse gas concentrations and a warming world.

However, there are considerable uncertainties about the size of a variety of natural and anthropogenic radiative forcing mechanisms. Patterns of cloud cover are critically important in determining climate change and remain one of the sources of greatest uncertainty in climate models. The role of sulphates and other anthropogenic aerosols is also critical and relatively poorly understood. Changes in solar activity may also explain some of the recently observed warming but few, if any, scientists, politicians or activists publicly doubt the human influence on the climate – natural climate variability alone does not explain the warming trend. Overall, according to the IPCC, the 'best agreement between model simulations and observations over the last 140 years has been found when anthropogenic and natural factors are combined'. In other words, we cannot explain what is happening to the climate without taking into account the human perturbation on the carbon cycle.

## 2.3 What is the scale of future climate change?

Over the period 1990–2100, the IPCC TAR projects that globally averaged mean surface temperature will increase between 1.4 and 5.8 °C. This is a huge range and has serious implications for political perceptions of risk. The lower end of the projection is in line with the maximum variation observed in the last 10,000 years. At the upper end, an increase of over 6 degrees in the period 1750–2100 would be highly significant and on a scale not seen since the last glacial/interglacial transition. Here uncertainties are beginning to compound the IPCC's scientific assessment. A critical source of uncertainty concerns future greenhouse gas emission levels. These in turn depend on trends in population, economic growth and technology **(Reddish, 2003)**. Uncertainty abounds in the prediction of all three factors. Estimates of annual emissions in 2100 (in the absence of climate response policies) gathered by the IPCC from the literature range from close to zero to 60 gigatonnes  of carbon per year (GtC/year) compared with 6.3 GtC in 2000 (Figure 5.5).

$1 \text{ Gt} = 10^9 \text{ tonnes}$

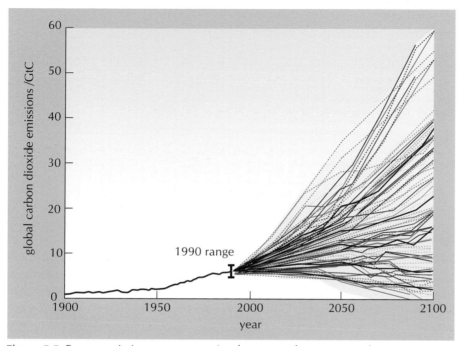

Figure 5.5  Future emissions are uncertain: the range of estimates in the IPCC TAR.

There is a broad range of projections. Even if we had a crystal ball and it accurately predicted our cumulative greenhouse gas emissions into the future, we still would not be able to accurately predict the magnitude of climate change. There are still significant gaps and uncertainties in our knowledge of the chain of cause and effect that links human activity ultimately with climate impacts (Figure 5.6). You may be surprised to discover that the 95 per cent confidence interval for our best guess as to how much warming will occur as a result of a doubling in $CO_2$ equivalent concentrations is between 1.5 and 4.5 °C. Once again, this is a wide range for policy makers to take into account.

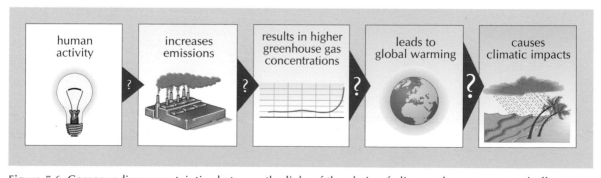

Figure 5.6  Compounding uncertainties between the links of the chain of climate change cause and effect.

## Activity 5.1

The IPCC's own scenario analysis in the TAR places estimates of emissions in 2100 from about 6 to 25 GtC per year. How easily can you 'get your head round' large unfamiliar numbers like that? What does it mean to you? Let's try to add some meaning by translating the uncertainty into something more visual, that you might have a better sense or 'feel for'.

Carbon dioxide emissions in 2000 were 6.4 GtC of which the USA accounted for approximately 1.5 GtC. How many additional 'US economies' worth of emissions do scientists believe we will generate by 2100: (a) at the lower end and (b) at the upper end of the range of uncertainty?

### Comment

The range of uncertainty is 6 to 25 GtC or 19 GtC, so the answers will be:

(a) none additional at the lower end; 6 GtC is approximately today's level of 6.4 GtC;

or

(b) twelve additional at the upper end; they will have grown to 25 GtC, an extra 19 GtC, equivalent to an extra 12 year 2000 US economies (of approximately 1.5 GtC).

Given the range of uncertainty is approximately 12 times the emissions of the USA (the single largest emitter of greenhouse gases), can you see that whether or not the USA halved or even doubled its emissions in the future, it still leaves a broad range of uncertainty for policy makers to grapple with.

## 2.4 What are the risks, consequences and costs associated with future climate change?

The link between temperature and physical, biological and economic impacts is extremely complex and uncertain. According to the IPCC, an increase in GMST of 1.4 °C (their lowest projection) presents the following risks (discussed in **Blackmore and Barratt, 2003,** p.271, Figure 6.25):

- threats to some unique and threatened systems
- some increase in risks from extreme climate events
- mixed positive and negative global economic impacts; the majority of people adversely affected (we could add here that least developed countries, who are often most vulnerable to climate change, are already losers)
- very low risks from future large-scale discontinuities (for example, a complete shutdown of the Gulf Stream or the disintegration of the West Antarctic Ice Sheet).

According to the IPCC, an increase of 5.8 °C (their highest projection) in GMST presents the following risks:

- threats to many unique and threatened systems
- large increase in risks of extreme climate events
- net negative global economic impacts for most regions (both developed and developing are now clear losers)
- higher risks from future large-scale discontinuities.

In other words, the hazards of climate change increase as GMST increases. This seems intuitive. For a 1.4 °C warming, the risks are spread unevenly across different countries and systems. Those countries and ecosystems that are already sensitive and vulnerable to climate change face higher risks. For a small degree of warming, the net balance of gains and losses due to climate change is negative only in some regions (often the poorest). There are clearly some winners and some losers. For an increase of about 6 degrees, the risks are more comprehensively and evenly distributed across regions and systems. The net balance of gains and losses due to climate change results in few, if any, winners. In other words, for a significant warming (greater than, say, 4 °C), there are many losers and the losses are significant.

## Summary

- Our scientific understanding of climate change is improving rapidly but it is far from complete and uncertainty is pervasive.
- There is little doubt that the climate is changing, as a result of an increase in the concentration of greenhouse gases and global warming, and that human activity is largely responsible for the overall trends we are now detecting.
- At the lower end of the predicted changes, the developing countries are more vulnerable to adverse effects (while some countries may benefit). The high end of predicted changes in increased global mean surface temperature (GMST) by the end of this century represents a rate of climate change not seen since the last glacial/interglacial transition. At this higher end all countries will feel adverse effects.
- Future risks from climate change are highly uncertain.

# 3   Uncertain responses

We have noted in the previous section that political decisions about how best to respond to climate change involve varying degrees and types of scientific uncertainty. The principal sources of uncertainty include: gaps in our knowledge of climate system dynamics; the magnitude of future climate change (this is uncertain because of the uncertainty in how humans may respond); and future responses of natural and human systems to climate impacts.

Thus, the human response to climate change is potentially a powerful negative feedback in the climate system, yet here uncertainty is inherent and all-encompassing. What can humans do to manage the risks associated with climate change? We have three options:

do little or nothing (inactive response)

See also the discussion on inactive responses in the Introduction to this book.

- *Option 1    **Do little or nothing (inactive response)***
  We can choose to continue on our 'business as usual' global emissions pathway and simply wait and see: (a) if emissions actually grow this way; and (b) the scale of damages that unfold. This wait-and-see approach means that at any future stage the decision can always be taken to adapt and/or mitigate (Options 2 and 3 below).

adaptation (reactive and proactive responses)

See also the discussion on reactive and proactive responses in the Introduction to this book.

- *Option 2    **Adaptation (reactive and proactive responses)***
  We can choose to adapt. Adaptation to climate change (discussed in section 3.1 below) is any kind of adjustment in response to actual or expected climate change. The climate system is complex and highly inertial – changes in some systems take several hundred years to work themselves through. So, even if greenhouse gas emissions were to fall suddenly and dramatically today, the earth's climate would continue to change for centuries. Greenhouse gas concentrations would continue to rise, the mean surface temperature would continue to increase, the average sea level would continue to rise and many other climatic impacts would still occur. Building houses on stilts in flood plains, or not building them there at all, building sea walls or irrigation systems, developing and sowing genetically modified drought and saline resistant crops, and making anti-malarial drugs cheaper and more widely available are examples of adaptations to climate change (Figure 5.7).

**Figure 5.7**  Adaptation options include: irrigation technologies; drought and saline resistant crops; and drugs to combat the spread of tropical diseases such as malaria.

- *Option 3   **Mitigation (proactive responses)***
  We can choose to mitigate. Mitigation (discussed in section 3.2 below) is any human intervention to reduce the sources or enhance the sinks of greenhouse gases. Generating electricity from renewable sources of energy instead of fossil fuels and saving energy wherever possible are examples of mitigation (Figure 5.8). Planting forests to sequester carbon in the form of woody biomass is another example (Figure 5.8). (In other contexts the word 'mitigation' is also used in the sense of 'mitigating adverse climate impacts', that is, in the context of adaptation. In this chapter I will only use mitigation in the sense of greenhouse gas reduction.)

mitigation (proactive response)

**Figure 5.8** Mitigation options include renewable energies, such as (anti-clockwise) wind, hydrogen and solar as well as planting forests to sequester atmospheric carbon.

## 3.1 Uncertainty in adaptive responses to climate change

The living world's response to climate change is reactive and there are clear signs that animals and plants have begun responding to climate change (Figure 5.3). In the UK there is evidence that spring is coming two weeks earlier than it would have done 30–50 years ago and autumn about a week later. These responses are indicators of climate change. Plants and animals cannot do anything other than adapt 'naturally'; humans, however, can also react proactively through the ability to plan ahead.

In the case of adaptation in natural and human systems, we can usefully distinguish between anticipatory and reactive types of adaptation (Table 5.1):

anticipatory adaptation (proactive)

reactive adaptation

- **Anticipatory adaptation (proactive)** takes place before the impacts are apparent and is the result of a deliberate decision-making process.
- **Reactive adaptation** takes place after the impacts of climate change have been felt.

**Table 5.1**    Examples of anticipatory and reactive adaptations to climate change

|  | Reactive | Anticipatory |
|---|---|---|
| **Natural systems** | Phenological changes, shifts in the range of distribution of species, community shifts, changes in ecosystem composition and migration patterns | Natural systems do not anticipate |
| **Human systems** | Changes in agricultural practices Changes in insurance premiums Purchase of air conditioning Compensation Clean up | *Autonomous (individual /societal changes):* Purchase of flood insurance Installation of air conditioning *Planned (government responses):* Early warning systems New building codes, design standards Building sea walls Managed coastal retreat |

We are currently tracking reactive adaptations in both natural and human systems as indicators of climate change. It is uncertain exactly how natural systems will respond to the climate of the twenty-first century. It is also very uncertain how human systems will adapt to climate change (in either anticipatory or reactive terms). In part, human adaptation depends on the evolution of the understanding of scientific uncertainties and how we approach uncertainty and risk in our frameworks of political decision making.

Human systems are capable of significant autonomous adaptation to climate variability. We see this throughout the history of human civilization and around the world today. Planned interventions to reduce sensitivity to climate variation can also be highly effective (e.g. flood control systems). How much of the anticipated impacts of climate change can human systems cope with and at what cost? We really do not know.

These uncertainties are a major reason why integrated assessment models of climate change – designed ultimately to weigh up the costs and benefits of different courses of action in dealing with climate risks – are not yet capable of providing meaningful estimates of the benefits of different policy approaches. As a result, much of the political discourse about climate change focuses on the costs of mitigation; mitigation is not more important than adaptation, but it does involve energy-economic models we feel comfortable with, and we do not yet have equivalent climate impact models.

Many couplings and feedbacks in the climate system have long timescales. The link between mitigation and impacts is therefore very inertial. Reducing greenhouse gas emissions by whatever amount will not prevent significant climate change during the rest of the century. Political decisions to commit resources to assist vulnerable regions to adapt in anticipation of these changes are therefore critically important in limiting the damage to sensitive and vulnerable ecological and socio-economic systems. Sea level rise, for example, is fairly insensitive to whatever political agreements are reached regarding future mitigation targets in the Kyoto Protocol beyond 2012 (Chapter Three). Climate models clearly show that under any scenario, expected sea level rise by 2100 is approximately ±4 cm. It is not much consolation to a low-lying country such as Tuvalu (maximum height above sea level approximately 5 metres) that whatever the international climate response, sea level rise by 2100 will be at least 40 cm.

In summary, both anticipatory and reactive adaptation responses are critical parts of the overall climate change policy equation. Policy decisions must take account of uncertainty in reactive responses in natural systems and reactive and anticipatory responses in human systems including uncertainties in the effectiveness of planned measures to limit climate damages.

## 3.2  Uncertainty in mitigation responses to climate change

Since the first major global oil crisis in 1973/1974, governments of developed economies have pursued energy conservation and energy efficiency policies. In the years since, their interest and enthusiasm has waxed and waned with the oil price (Figure 5.9).

Over time, there have been several attempts to reduce energy consumption in various sectors including housing, transport and industry. Measures have included regulations, economic incentives and voluntary agreements (Chapter Three). Do policies and measures to mitigate greenhouse gas emissions ultimately work? There is a great deal of controversy surrounding the evaluation of the effectiveness of energy saving measures or climate mitigation technologies. It is a complex and highly uncertain business for two major reasons: it is difficult to separate the impact from the background noise and the 'rebound effect'.

To distinguish the true impact of the policy, project or technology separate from all kinds of background noise in the socio-economic system, is often difficult because the sources of anthropogenic greenhouse gas emissions and human interference with the carbon cycle (i.e. the land use sink) are so comprehensive and pervasive that almost all human activity is associated with emissions.

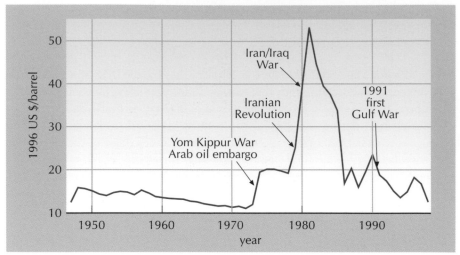

**Figure 5.9** Political enthusiasm for renewable energy and energy efficient technologies went up and then down with the global oil price.

Many factors determine energy consumption and greenhouse gas emissions – not simply government policies. Emissions are the result of a complex web of interactions between changes in population, economic activity and technological factors, among others (**Reddish, 2003**). Forecasting greenhouse gas emissions is a highly uncertain business even without the added complication of having to best guess how humans might respond to the threat of climate change. The future emissions pathway without accounting for mitigation policies and measures is called the 'emissions baseline'. The global emissions baseline is highly uncertain and yet this is a key factor behind the broad range of future temperature projections from the IPCC (**Blackmore and Barratt, 2003**).

It is a fact of life that things consistently improve in a technological and design sense. Energy systems, technologies and services have become more and more energy efficient as we have developed economically. A new car today is significantly more efficient than a car built in the early 1970s. Today's new houses, heating systems, light bulbs, refrigerators, TVs and computers are all generally more energy efficient than their predecessors (Figure 5.10).

If we lived the way we do today with the technologies we had in the 1970s, our energy consumption and greenhouse gas emissions would be significantly higher – perhaps double. Some of these improvements are definitely linked to specific policies adopted by governments, but this is against a background of general improvements to things through enhanced human knowledge, skill and creativity and without the helping hand of government intervention. Your next journey may or may not be influenced by government incentives to reduce greenhouse gas emissions. If the government influenced your choice, you are part of the effectiveness of the policy. If it did not and you, for example, took the bus or the train anyway regardless of government efforts to lure you out of your car, you are what environmental economists call a 'free-rider' and you may be

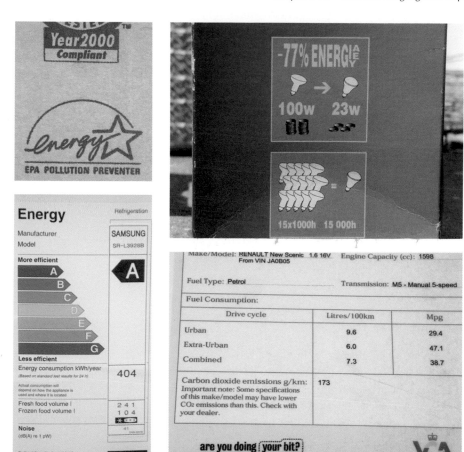

**Figure 5.10**  Energy efficiency has improved dramatically, as these energy product labels on (clockwise from top left) computers, light bulbs, cars and refrigerators indicate.

included in the policy count all the same. Separating out policy impacts from background noise and free-riders is difficult and keeps some energy economists very busy.

Economists call the gradual and predictable natural decoupling of energy consumption from economic growth the 'autonomous energy efficiency improvement' (Figure 5.11). Energy models always include a 0.5–1 per cent a year energy consumption 'fudge factor' to account for the technology improvements that dependably occur independent of climate or energy policies. Environmental optimists (see Chapter Six) tend to emphasize such autonomous technological improvements. Things, they say, will improve whatever.

The second reason we noted above for uncertainty in the assessment of the effectiveness of mitigation policies is related to what economists call the 'rebound effect'. Environmental pessimists tend to emphasize that any gains as a result of improvements in energy efficiency are generally equalled or outweighed by changes in behaviour towards more energy intensive lifestyles. For example,

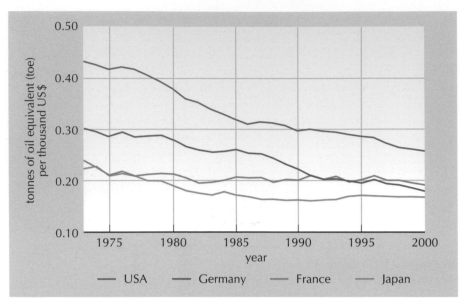

**Figure 5.11** Trends in the energy intensity of various countries (the ratio of total energy consumption to GDP).

saving on a heating bill by buying a new – energy efficient – condensing boiler may simply be freeing up the money to spend on an energy intensive 'treat' such as a cheap weekend flight. As another example, today's new cars might be more efficient, but there are many more of them and on average we use them a great deal more than we did in the 1970s.

The distinction between planned intervention and autonomous improvements in energy efficiency partly explains different political perspectives on sustainable development issues in general. Pessimists might tend to believe that relying on the natural rate of improvement would be a high-risk strategy. Instead, they argue, there needs to be more purposeful intervention on the part of governments (usually perceived as those who need to take the lead) together with other stakeholders. Optimists, sometimes referred to as ecological modernizers (see Chapter Six), tend to reflect on how things generally seem to change for the better – regardless of who is 'in control' or what decisions are, or are not, taken. At the very least, they tend to believe things will improve by themselves, and often believe there is a good chance we can reduce emissions without modifying our behaviour or lifestyles (the so-called 'win–win' approach; see Chapter Six).

There is no doubt among economists that stabilizing greenhouse gas concentrations is going to cost a significant portion of global domestic product – as much as several per cent points per year by the end of the century. The IPCC have not yet provided policy makers with an estimate of the costs of climate stabilization: the numbers are simply too uncertain. However, the IPCC did review a range of studies of stabilization at 550 ppm and found costs in the range of US\$ 1 to 18 trillion (global GDP in 2000 was approximately US\$ 34 trillion). The costs depend on many factors, but in particular on whether or not there is international (global) emissions trading (see below).

For many economies, mitigating the first few million tonnes of greenhouse gas emissions here or there will be cheap and easy. Frequently, there are 'win–win' opportunities and 'secondary benefits' associated with different policies or measures. For example, cleaning up local air pollution may reduce greenhouse gas emissions at the same time. These early days of climate politics are partly about the obvious and cost-effective opportunities out there just waiting to be grasped. You can think of it as the economic equivalent of euros lying on the ground just waiting to be picked up by consumers (Figure 5.12). This is a widely used analogy in energy economics. However, there is a limited supply of 'win–win' opportunities available and sooner or later climate mitigation policies are going to start costing real money, involve lifestyle changes and ultimately – if they are indeed working – win or lose political votes.

Figure 5.12 Some economists believe that early climate mitigation measures also provide secondary benefits – like euros 'for free'.

## 3.3  Finding the right balance in future decision making

Some of the international political disagreements about climate change are fuelled by different perspectives on climate economics. In a nutshell, leading economists argue about the shape of the most cost-effective global emissions reduction pathways. All of them agree on the basis of current science that there are clear benefits from limiting greenhouse gas emissions. They disagree on the relative benefits and costs associated with earlier action versus delay. As a tool, cost–benefit analysis applied to climate change is very complex and uncertain, and is highly controversial (**Burgess, 2003**).

Sources of disagreement among economists on the optimal emissions reduction pathway include:

- the cost–benefit associated with delaying action and waiting for better science (climate change mitigation as sequential decision making under uncertainty – particularly as many of the impacts and effects of climate change are uncertain; some economists argue that we can afford to wait until the science improves);

- the nature and treatment of 'discount rates' (a high discount rate makes future benefits from an investment made today 'look' small, a lower discount rate makes the benefits 'look' greater), as illustrated in Figure 5.13;

See Chapters One and Six for a discussion of discounting.

- assumptions about the nature of 'autonomous technical change' built into their models (the degree to which things will get better independently of any action on the part of governments) as well as the extent to which we will see technological 'spill over' effects (the migration of new technologies to developing countries as a result of western leadership in tackling the issue);

- assumptions about the capacity of natural and human systems to adapt autonomously to climate impacts.

The political machinations of the international climate negotiations can be understood as an evolving debate about the potential risks of inaction versus

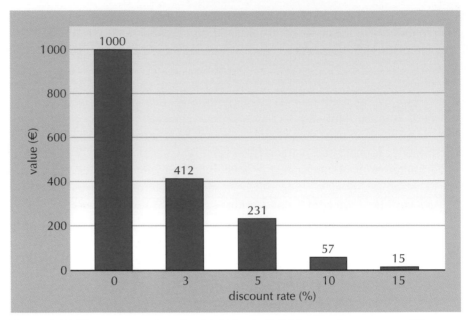

**Figure 5.13** The present value of a €1,000 windfall 30 years from now at different discount rates; for example at 15 per cent discount rate,€1,000 30 years from now is worth the same as €15 in your hand today.

action to control greenhouse gas emissions. Politicians are being asked to make a complex cost–benefit decision. The costs are the political and economic implications of dramatic cuts in greenhouse gas emissions. The benefits are the avoided damages as a result of taking action. Both are highly uncertain. Climate change is only one of a number of global environment and development challenges and politicians have limited patience and resources to spend on the issue. Improving access to education, clean water, sanitation, health and energy services are all equally, if not more, important international development goals.

So far the international political response has focused on mitigation. As we will see, the Kyoto Protocol is a mitigation protocol. That is, by far the main body of the text and the volumes of small print that have since been agreed in support of it deal with detailed rules about greenhouse gas reduction targets.

Mitigation, however, is only half the issue. In the years to come, political interest in the topic of adaptation will continue to grow and we are likely to witness another legally binding climate instrument – an adaptation protocol. Such a protocol would focus on compensation to victims of climate change (often the least developed countries in the world) as well as provide finance and technical assistance to help vulnerable countries cope with and adapt to climate impacts. The overall challenge for the global political community is to find the right balance of emphasis between mitigation and adaptation in their chosen response strategy.

**Activity 5.2**

Match the five statements (A to E) to the following four perspectives on priorities for climate change responses:

A   Energy technologies just get more and more energy efficient.

B   Much of the science around impacts and potential costs and benefits is highly uncertain.

C   There is a risk of irreversible and potentially catastrophic changes in the climate.

D   Because of inflation, when you discount the costs, it turns out that it is cheaper to deal with the problem tomorrow.

E   Measures that help adapt to future climate change (such as improved irrigation systems), frequently provide immediate spin-offs for sustainable development.

    1   Wait and see               3   Mitigate later

    2   Mitigate sooner rather than later    4   Adapt now

You can find the answer at the end of the chapter.

## Summary

- There are three policy options to respond to climate change: wait and see; adapt to climate impacts; and mitigate greenhouse gas emissions.

- There are inherent uncertainties and risks associated with the effectiveness of all three options and this is a major source of political uncertainty and conflict. One perspective is that scientific uncertainty is an argument in favour of a wait-and-see approach. Another is that inertia in the climate system and the inevitability of climate damages demands urgent action on adaptation, particularly for the least developed countries who are often the most vulnerable. Another is that delaying stringent mitigation prevents stabilization at lower concentration levels and simply shifts the risk to future generations.

- A great deal of the political conflict related to climate change can be explained by vested group interests and their different perspectives on the balance of how resources should be spent between adaptation and mitigation.

# 4   Legal dimensions of global environmental governance

The management of risks posed by global environmental change requires coordinated international responses. In recent years we have witnessed the birth of a growing number of international legal instruments designed to protect the global environment. The international legal dimension of climate change adds a further level of complexity in understanding the interplay between scientific and political uncertainties.

Just as there are national laws designed to protect the environment, for example, by prosecuting factories or farmers that illegally discharge waste into rivers, or vehicle owners who fail to maintain their vehicles, there are also various international laws designed to protect the environment. However, there are significant differences in the contexts of national and international law. We frequently come across examples in the news of differences in national systems of law (different procedures in the justice system, different rules, fines and punishments). At the international level, there are inherent difficulties in monitoring implementation and compliance with international laws, as well as in designing enforcement mechanisms in the case of non-compliance. Often the interaction between the new international law and existing national legal customs must be taken into consideration.

Before we consider the UNFCCC and its Kyoto Protocol in more detail, let us take a step back and consider the history and context of environmental responses in the form of international agreements. This will help to explain some of the unfolding political dynamics of the climate issue.

## 4.1 A brief history in three phases

By now, you may appreciate that any framework of international law designed to manage the risks of climate change would need to be elaborate and highly technical to cope with the characteristics of the climate challenge (Box 5.1). International cooperation on technical matters has a long history. The earliest intergovernmental organizations were based on revolutions in communication. The world's oldest intergovernmental organization is the International Tele-communications Union (ITU) with a history dating back to 1865 and the development of the first telegraph systems (Figure 5.14).

**Figure 5.14** The ITU headquarters, Geneva. What sorts of institutions might be established to address the challenges of global environmental change – and will they last as long as the ITU?

A decade later the Universal Postal Union (UPU) was established in October 1874 when representatives of 22 countries signed the Treaty of Berne. The Treaty unified a conflicting maze of postal services and regulations into a single postal territory for the exchange of letter-post items. These were the very early days of technical cooperation among states at the international level.

Today there are hundreds of international organizations and agreements on issues ranging from labour, to peace and security and technical standards, social and economic integration, to trade and environmental protection. Many were established in the period up to the late 1960s following the Second World War. The growth of international organizations and legal instruments in the field of environmental protection is relatively recent.

The first phase of growth began in the 1950s with a focus on technical issues usually connected in some way or other with trade objectives. A second phase of growth began in the early 1970s triggered by the original 'Earth Summit', the 1972 United Nations Conference on the Human Environment (UNCHE), held in Stockholm. This phase lasted until the early 1990s and included the 1985 Vienna

(a)

(c)                                          (b)

Figure 5.15 Images of international cooperation: (a) United Nations Conference on the Human Environment (UNCHE), Stockholm, 1972; (b) United Nations Conference on Environment and Development ('Earth Summit'), Rio de Janeiro, 1992; and (c) World Summit on Sustainable Development, Johannesburg, 2002.

Convention for the Protection of the Ozone Layer and its Montreal Protocol. A third phase began at the 1992 United Nations Conference on Environment and Development (UNCED) in Rio de Janeiro (often referred to as the 'Earth Summit').  (See Figure 5.15 and Box 5.2.)

## Box 5.2  Environmental governance in three phases

### First Phase    After the Second World War

| | |
|---|---|
| 1951 | The World Meteorological Organization (with the UNEP, established the IPCC in 1988) becomes a specialized UN agency |
| 1954 | Convention on the Prevention of Pollution of the Sea by Oil |
| 1956 | International Finance Corporation |
| 1957 | International Atomic Energy Agency |
| 1959 | The Antarctic Treaty signed (came into force in 1961) |
| 1960 | International Development Association |
| 1961 | Organisation for Economic Cooperation and Development |
| 1963 | World Food Programme |
| 1964 | United Nations Conference on Trade and Development (UNCTAD) established to 'maximise the trade and development opportunities of developing countries and assist them in facing the challenges arising from globalisation and to integrate into the world economy on a more equitable basis'. |

### Second Phase    From Stockholm to Rio

| 1972 | Declaration of the United Nations Conference on the Human Environment, Stockholm |
|------|---------------------------------------------------------------------------------|
| 1972 | UNEP established |
| 1972 | Convention Concerning the Protection of the World Cultural and Natural Heritage |
| 1975 | Convention on International Trade in Endangered Species of Wild Fauna and Flora enters into force |
| 1985 | Vienna Convention for the Protection of the Ozone Layer |
| 1987 | Montreal Protocol on Substances that Deplete the Ozone Layer, (in the same year the influential Brundtland Report, *Our Common Future*, was published, the source of the most widely quoted definition of sustainable development). |
| 1991 | Convention on Environmental Impact Assessment in a Transboundary Context (enters into force 1997) |

### Third Phase    From Rio to Johannesburg

| 1992 | United Nations Conference on Environment and Development in Rio de Janeiro produces the Rio Declaration on Environment and Development, Agenda 21, and the three 'Rio conventions' – United Nations Convention on Climate Change (UNFCCC), Convention on Biological Diversity (CBD) and Convention to Combat Desertification (CCD). Rio also generated a set of Forest Principles. |
|------|---------------------------------------------------------------------------------|
| 1994 | CBD enters into force; UNCCC enters into force; Programme of Action of the International Conference on Population and Development |
| 1995 | World Summit for Social Development |
| 1996 | Report of the World Food Summit; Convention to Combat Desertification enters into force |
| 1997 | Kyoto Protocol is adopted |
| 2000 | The Cartagena Protocol to the CBD is adopted. The Protocol seeks to protect biological diversity from the potential risks posed by living modified organisms resulting from modern biotechnology. |
| 2002 | World Summit on Sustainable Development in Johannesburg (Rio+10) produces the Johannesburg Declaration on Sustainable Development, The Johannesburg Plan of Implementation and several hundred 'partnerships' in sustainable development (these are essentially voluntary agreements). |

(The image on the contents page of this chapter (p.186) dramatizes environmental responses on a global stage: it shows the opening ceremony at the Johannesburg summit, 2000, in the Ubuntu Village.)

## 4.2 Soft and hard multilateral environmental agreements

Today's international environmental legal space is relatively crowded with a dense patchwork of over 200 international environmental agreements. They are often referred to as **multilateral environmental agreements (MEAs)**. However, they do not all have the same legal status. Some are more important – in legal terms, 'binding' – than others. As governments generally act more

multilateral environmental agreements (MEAs)

effectively when they are forced to do so, legally binding international agreements can be more effective in reducing global environmental risks.

International legal instruments can be helpfully (for our purposes) classified as either 'soft' or 'hard' (Hurrel and Kingsbury, 1992).

- **_Soft international environmental law_**

  These are non-legally binding resolutions, declarations, frameworks of good practice or other forms of agreement. They indicate what states _should_ do rather than _must_ do. However, although they are non-binding, they are not without legal effect. The 1972 Declaration of Principles on the Human Environment and its direct descendants, the 1992 Rio Declaration on Environment and Development and the 2002 Johannesburg Declaration on Sustainable Development are soft in international legal terms. This is why if you read them, you may be surprised to find they contain reasonably frank, clear and sensible aims and goals. They frequently appear to rise above international politics. It is much easier for government representatives to make promises when there are no consequences for breaking them.

  Principles are examples of soft law. States generally implement international environmental principles at their own pace or not at all and treat the principles as goals. The precautionary approach as embodied in Principle 15 of the 1992 Rio Declaration is a keystone in the framework of international environmental law. It is where scientific uncertainty clashes head-on with political perceptions of environmental risks. Principle 15 of the Rio Declaration on Environment and Development states: 'In order to protect the environment, the precautionary approach shall be widely applied by States according to their capabilities. Where there are threats of serious or irreversible damage, lack of full scientific certainty shall not be used as a reason for postponing cost-effective measures to prevent environmental degradation.'

  The Rio declaration applies the precautionary principle/approach to all environmental issues. It prevents policy makers and politicians from hiding behind the uncertainties inherent in the conclusions of scientific research and forces them to take decisions on the basis of probabilities. At its strongest, it changes the burden of proof to the polluter to prove that an activity is _not_ harmful. At its weakest, action to prevent damage must be considered 'cost-effective' in the widest economic and political sense – cynics may observe this is a significant get-out clause. This framework of soft law allows states to assume obligations they might not have otherwise.

  Soft laws can and do grow legal teeth. The ongoing discussions on the application of the precautionary principle in the WTO negotiations is an example of such progression. Soft laws can harden into treaties. When such treaties impose mandatory obligations they are classed as 'hard'.

- **_Hard international environmental law_**

  These are legally binding treaties, conventions or protocols with consequences for non-compliance. The stronger the monitoring, compliance

soft international environmental law

See also **Blowers and Smith (2003)** for a discussion of the precautionary principle.

hard international environmental law

and enforcement procedures, the 'harder' the law is. Examples of hard international law include: the UNFCCC and its Kyoto Protocol (climate), the Montreal Protocol (ozone depleting substances), and the Convention on Biological Diversity and its Cartagena Protocol (biodiversity). All have slightly different compliance and enforcement mechanisms. The compliance mechanism of the Kyoto Protocol is particularly innovative and has been described as unprecedented in international legal terms.

Despite the explosion of MEAs, we are in the early days of establishing appropriate systems of global governance on environmental issues. You may be surprised to discover that despite the creation of over 200 MEAs, there is no international court to rule on them. Ultimately, much of enforcement power behind the framework of international environmental law relies on the goodwill and citizenship of nation states to ensure their own compliance.

The proliferation of MEAs presents a major challenge for international environmental governance. At the same time, the process of 'globalization' – the growing interconnectedness among states, markets and cultures – continues. Globalization is itself partly responsible for global environmental change on the scale we now see. Key themes at the 2002 World Summit on Sustainable Development included synergies among MEAs, and specifically the Rio Conventions, as well as the linkages between sustainable development and globalization (Figure 5.16).

**Figure 5.16** Synergy means talking to each other: the three heads of the Rio Convention Secretariats (from left to right) Hamdallah Zedan (CBD), Joke Waller-Hunter (UNFCCC) and Hama Arba Diallo (CCD) present a united front to the world's press.

## Summary

- International technical cooperation began with the telecommunications and postal revolutions and continued throughout the twentieth century in various waves of development.

- Global environmental change has generated international legal responses. These often take the form of new soft or hard legal instruments. There are now more than 200 MEAs. Not all agreements have the same status and different states are in different stages of implementing their responsibilities for different agreements.

- Differences of opinion among states as to the status and importance of various international environmental legal principles and MEAs is another source of political uncertainty in understanding responses to global environmental change.

- The institutions necessary for effective global environmental governance have yet to be established.

# 5   The UNFCCC and its Kyoto Protocol

Inside the ongoing climate negotiations we get a glimpse of how human systems are struggling to cope with decision making about environmental change at a very high level and under conditions of extraordinary scientific and political uncertainty. The **United Nations Framework Convention on Climate Change (UNFCCC)** was formally opened for signature at the 1992 Earth Summit in Rio de Janeiro, and came into force in 1994. It is the central international agreement governing international efforts to combat climate change.

United Nations Framework Convention on Climate Change (UNFCCC)

The negotiations have come a long way in the relatively short time since 1992. The Convention's subsidiary bodies, institutions and mechanisms represent one of the most complex frameworks of intergovernmental environmental negotiation in existence. Official documents related to the process are voluminous (Figure 5.17) and many are highly technical.

Official 'delegates', vetted representatives of countries, international organizations, global NGOs and civil society, all struggle to swim in what they call the 'alphabet soup' of acronyms and technical climate jargon that engulfs them. Trying to make sense of it all is a daunting task for anyone. Even bona fide negotiators themselves take several meetings – amounting to years of time – to settle into the process satisfactorily. Frequently too scientific for the politicians and too political for the scientists, the negotiations continue to make the news and break new ground in the world of international environmental governance.

**Figure 5.17** Mountains of official documents at a UNFCCC Conference of the Parties (COP).

## 5.1 The UNFCCC

The UNFCCC is an example of a 'soft' MEA. The Convention itself does not have any legal teeth like, for example, the mechanism of the World Trade Organization

**(Taylor, 2003)**. However, as its name suggests, the Convention is a framework to guide the international response. It is a living document and can be amended at any time. One way of doing this is to introduce a Protocol to the Convention; the Kyoto Protocol is one such amendment, designed to ensure developed countries would begin mitigation in earnest by setting legally binding emission reduction targets (discussed below; see also Chapter Three).

The core of the framework of the climate convention consists of an overall objective, a set of principles and a series of commitments. The goal of the convention is stated in Article 2, perhaps the most important article in the Convention:

> The ultimate objective of this Convention and any related legal instruments that the Conference of the Parties may adopt is to achieve ... stabilization of greenhouse gas concentrations in the atmosphere at a level that would prevent dangerous anthropogenic (human related) interference with the climate system. Such a level should be achieved within a time-frame sufficient to allow ecosystems to adapt naturally to climate change, to ensure that food production is not threatened and to enable economic development to proceed in a sustainable manner.

By bringing the UNFCCC into force, governments have therefore set about what is effectively environmental management on a global scale. Such an environmental aim is truly unprecedented. The ultimate goal is clear in the sense that it seeks to stabilize greenhouse gas concentrations. However, it is a matter for a combination of expert and political judgement as to what constitutes 'dangerous' interference with the climate system. What level of greenhouse gas concentrations represents a safe level? What level of climate change is dangerous? Are the risks associated with a 1.4 °C increase in mean temperature dangerous? The answers are by no means simple. International negotiators will continue to argue over them for years to come.

### Activity 5.3

Study Article 2 above and the Principle 15 of the Rio Declaration (Section 4.2, under 'Soft environmental international law', p.209) carefully. To what extent do you think Article 2 embodies a strong or weak interpretation of the precautionary principle?

You can find the answer at the end of the chapter.

When a government ratifies an environmental treaty it becomes one of the 'parties' to it.

The UNFCCC does include a number of principles regarding who should act and how they should act to manage climate change. The principles (Article 3) include those established in the Rio Declaration as well as others and cover such things as: equity; common but differentiated responsibilities; leadership; specific needs and special circumstances of developing country parties; the precautionary principle; and the right to sustainable development. (See Box 5.3.)

## Box 5.3  Key principles underpinning the UNFCCC

1   The Parties should protect the climate system for the benefit of present and future generations of humankind, on the basis of equity and in accordance with their common but differentiated responsibilities and respective capabilities. Accordingly, the developed country Parties should take the lead in combating climate change and the adverse effects thereof.

2   The specific needs and special circumstances of developing country Parties, especially those that are particularly vulnerable to the adverse effects of climate change, and of those Parties, especially developing country Parties, that would have to bear a disproportionate or abnormal burden under the Convention, should be given full consideration.

3   The Parties should take precautionary measures to anticipate, prevent or minimize the causes of climate change and mitigate its adverse effects. Where there are threats of serious or irreversible damage, lack of full scientific certainty should not be used as a reason for postponing such measures, taking into account that policies and measures to deal with climate change should be cost-effective so as to ensure global benefits at the lowest possible cost. To achieve this, such policies and measures should take into account different socio-economic contexts, be comprehensive, cover all relevant sources, sinks and reservoirs of greenhouse gases and adaptation, and comprise all economic sectors. Efforts to address climate change may be carried out cooperatively by interested Parties.

4   The Parties have a right to, and should, promote sustainable development. Policies and measures to protect the climate system against human-induced change should be appropriate for the specific conditions of each Party and should be integrated with national development programmes, taking into account that economic development is essential for adopting measures to address climate change.

5   The Parties should cooperate to promote a supportive and open international economic system that would lead to sustainable economic growth and development in all Parties, particularly developing country Parties, thus further enabling them to address the problems of climate change. Measures taken to combat climate change, including unilateral ones, should not constitute a means of arbitrary or unjustifiable discrimination or a disguised restriction on international trade.

(UNFCCC, Article 13)

The Convention also establishes a series of general promises or 'commitments' as they are called (Article 4). Some commitments apply to all countries, others concern just developed countries (Box 5.4).

The international politics of climate change are, like several other international economic, social and environmental issues, frequently portrayed in terms of the

'North versus South' dimension – a euphemism for developed versus developing countries. Dividing the world up into these two basic camps offers a starting point (albeit a crude one) for understanding some of the political tensions behind climate change (Figure 5.18). The text of the UNFCCC reinforces the North–South dynamic by placing different commitments upon developed countries, upon countries with economies in transition (EIT, countries of the former USSR and Eastern Europe) and upon developing countries (Figure 5.19 and Box 5.4).

**Figure 5.18** North–South tensions are real and tangible in international environmental politics and no more so than in the climate negotiations.

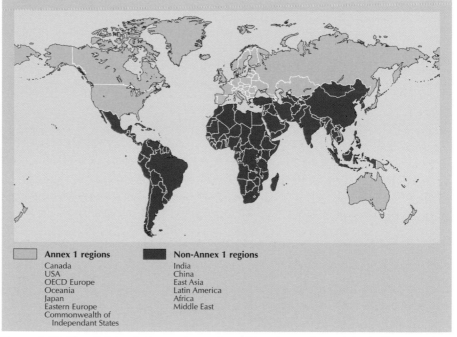

**Figure 5.19** The division between Annex I and Non-Annex I countries indicates a clear North–South divide.

## Box 5.4   UNFCCC commitments are differentiated 'North vs South'

*All Parties commit to:*

Implement measures to mitigate and adapt to climate change

Promote and cooperate in the development, application, diffusion including transfer of mitigation or adaptation technologies

Cooperate in preparing for adaptation to the impacts of climate change

*Developed country Parties commit to:*

Adopt national polices to reduce greenhouse gas emission and return these to earlier levels by the year 2000 (very few Parties achieved this)

Provide new and additional financial resources to meet the agreed full costs incurred by developing countries in meeting various commitments under the convention

Assist particularly vulnerable parties in meeting the costs of adaptation

Promote, facilitate and finance the transfer of, or access to, environmentally sound technologies to developing country parties

Special consideration is also given to economies in transition (former Soviet Union and Eastern European countries), and to other groups of vulnerable countries or systems (e.g. small island states and fossil fuel exporting countries).

(extracts from UNFCCC, Article 4)

Much of the ongoing politics of the climate negotiations centres on different interpretations of the principles and commitments outlined in Articles 2 and 3.

## 5.2   Behind the scenes at the UNFCCC

The UNFCCC is much more than a dry collection of legal paragraphs. It is a living agreement and a living process of engagement between states and other stakeholders. That engagement takes the form of large international meetings on a regular basis in different locations around the world. You may have wondered what actually happens at these meetings. They are, effectively, two-week periods of relatively unscripted theatre in the full glare of the world's media. The UNFCCC climate meetings known as Conferences of the Parties (COPs) are perhaps the largest such MEA gatherings in the UN system (Figure 5.20).

The UNFCCC process has been likened many times to a 'travelling circus'. Several thousand people turn out for a COP. Currently one COP is held each year. The number of participants varies from year to year, depending on the location and the agenda: 8000 turned out for Kyoto in 1997 (COP 3) while other COPs typically attract 4000 or so. Who are all these people? Most importantly, they include members of official delegations to the meetings. Richer countries take larger delegations (50–150 people), smaller developing countries have much

**Figure 5.20** Examples of UNFCCC COPs: (left) COP 3 in Kyoto, 1997, which had 8,000 participants; and (right) COP 8 in New Delhi, 2002, which had 5,000 participants.

smaller delegations, and often they can only afford to send the two paid places they officially get (even in the case of a relatively wealthy developing country such as Thailand, for example). The larger a country's delegation is, the more effectively it can participate. A UN member state can only exercise its power to support or block procedures if it has a representative in the negotiating room. For this reason developing countries typically cooperate as a regional block known as the 'G77 and China' and share delegates among the many plenary sessions, contact groups and side events in order to keep pace with the tactics of the Europeans, Russians and North Americans, among others.

While there are many international environmental agreements, for small or relatively poor countries there are in fact not that many negotiators. The few negotiators they have are overstretched and struggle to keep up with the growing number of international meetings covering many MEAs. One week they are off to a meeting on biosafety, the next to one on ozone depletion, the next to one on persistent organic compounds, and the next to one on climate change. COPs are not the place to resolve tricky political and technical disagreements – they are more theatre than substance. What happens therefore is that further technical 'subsidiary bodies' and technical processes are set up to report to the conventions. That then introduces another level of complexity and another set of international meetings, another seat with the developing country flag on it that needs to be filled. And so it goes on.

Global environmental governance is complex. It requires a supply of adequately trained experts and bureaucrats to engage in the process. They also need money to get them to the meetings and pay their hotel bills. At roughly US$5000 per delegate per meeting, the cost of participating in the global environmental governance process can very quickly dry up the resources of environment ministries in smaller poorer developing countries. It would only take two delegates attending ten meetings to spend the entire budget the UN has so far offered many of them to come up with vulnerability assessments and adaptation plans.

Besides official government representatives who are allowed to sit behind a national 'flag' (country name-plates are called 'flags') in the main plenary rooms at a COP (Figure 5.21), there are a large number of others representing international and intergovernmental organizations (IGOs) and various interest groups or non-governmental organizations (NGOs). There are various categories of NGOs present, including environmental (ENGOs) and business (business initiated NGOs, BINGOs). This diverse community performs a vital function. It is a chance for experts and activists to network, catch up and share information. Figure 5.22 shows one such activist (an OU graduate) promoting hydrogen energy in the COP 7 village in Marrakesh, Morocco in 2001.

Figure 5.22 Mike Koefman, hydrogen enthusiast, holding an electrolysis kit connected to a PV unit outside a media tent at UNFCCC COP 7 in Marrakesh, 2001.

Figure 5.21 The Plenary room at UNFCCC COP 8, New Delhi, 2002: IGOs sit at the back, countries sit at the front in alphabetical order.

Having flown everybody in (the usual way to travel for most participants), the attendees must be supplied with tonnes of prepared documentation on the business of the meeting. There are six official UN languages (Arabic, Chinese, English, French, Russian and Spanish) and each document must be made available in all six. The aim of most negotiations is to reach a compromise agreement on some new form of text, prepared in all languages during the sessions, based on the 'pre-session' documents made available for delegates to read (usually on the plane).

This itself is a huge logistical exercise. Can you imagine then what it takes to translate and distribute a new 'in-session' document, say one that has just been marked (bracketed) to denote unresolved disputes at 1.00 a.m. and needs to be reprinted, and available in six languages six hours later by 7a.m.? The document is sent electronically to a team standing by in Geneva, translated, checked for meaning, perhaps sent to the UN legal office in New York for clearance and then sent back and redistributed on the ground in time for breakfast briefings. There is not space here to describe all the other happenings: the security arrangements, media tents, problems delegates encounter simply finding each other, the circus of special events, the logistical problem of serving 4,000 lunches within the same hour or the intense mobile phone culture that has evolved at such meetings.

## 5.3  The Kyoto Protocol

Kyoto Protocol

In 1994 negotiations began on a new and tougher agreement to supplement the UNFCCC. This time negotiators were trying to agree something with teeth – a legally binding protocol to the UNFCCC. Three years of detailed negotiations produced the **Kyoto Protocol** in December 1997 (Figure 5.23). Most news stories about climate change mention it. What is new about the Kyoto Protocol is that it legally binds industrialized nations to collectively reduce their greenhouse gas emissions according to agreed timetables. Failure to comply with these targets will have serious financial as well as other consequences. While the Kyoto Protocol is said to be legally binding, the nature of the non-compliance regime is still reasonably unclear and will continue to evolve in the coming years, providing plenty of work for international lawyers.

**Figure 5.23** The Kyoto Protocol is adopted on 11 December 1997 at UNFCCC COP 3, Kyoto, Japan.

Industrialized countries as a group must reduce their greenhouse gas emissions by 5 per cent in the period 1990–2012. Future targets will be agreed on a rolling basis, until the objective of the convention is met. Individual countries agreed individual targets (see Table 3.3, Chapter Three).

**Activity 5.4**

Assuming that emissions from industrialized countries would have continued growing at an annual average rate of say 0.8 per cent per year in the period 1990–2012 in the absence of the Kyoto agreement, what is the real level of reduction that Kyoto represents if actually achieved?

You will find the answer at the end of the chapter.

Under the Kyoto Protocol, countries can meet their Kyoto targets in two distinct ways:

- *Domestic measures*
  These are measures a nation takes to reduce emissions at home. This could include for example reducing traffic emissions by boosting public transport, reducing electricity consumption in homes, offices, schools and factories through energy efficiency measures, and better design of all kinds of energy products, processes and systems (see Chapter Two)

- *Offshore measures*
  These measures provide a way for industrialized countries, under certain conditions, to trade emissions permits or to obtain credit for emission reductions that are achieved through overseas sustainable development projects.

The UNFCCC and the Kyoto Protocol do not include specific fiscal instruments such as pollution taxes, charges or other instruments, nor do they set technology standards such as appliance standards or introduce labelling schemes. Such measures are catered for at the relevant national or in some cases regional (e.g. the EU) levels.

A great deal of political ambiguity lies behind the formal face of the UNFCCC and its Kyoto Protocol. Officially agreed language often represents the lowest common denominator of international agreement. This generates sustained and often intense behind-the-scenes conflicts between nation states and other stakeholders.

## 5.4 Political conflicts around the principle of 'leadership'

The Convention makes it clear that the developed countries should take the lead in reducing greenhouse gas emissions. The UNFCCC text, agreed in 1992, requires so-called Annex I countries (OECD members, plus the countries of the former USSR and Eastern Europe, known as economies in transition, or EIT countries; see Figure 5.19) to return their emissions to 1990 levels by the year 2000.

By 2000, many developed countries had failed to live up to their promises, and several had spectacularly overshot their goal (Figure 5.24). However, by 2000, emissions from the Annex I group as a whole were down more than 5 per cent compared with 1990. The large reductions of emissions from the EIT countries in the early 1990s (sometimes given the name 'hot air' in the context of emissions trading) outweighed the large increases in emissions in several of the OECD countries. Hence, the Annex I group had in fact met its Kyoto target by the year 2000, twelve years ahead of schedule. If they keep emissions roughly at that level until 2012 the developed countries would have collectively kept their promise and demonstrated the leadership that developing countries have demanded. However, the OECD and EIT economies are expected to grow significantly in the period to 2012. To avoid rising energy consumption and greenhouse gas

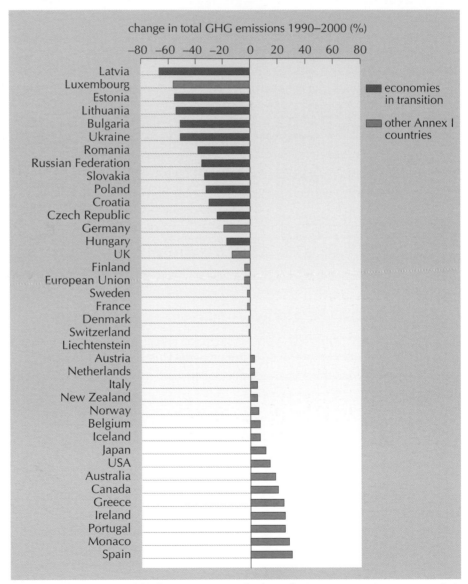

**Figure 5.24** Hot air and broken promises. Who kept their 1992 promise to return their emissions to 1990 levels by 2000 and who did not?
*Source:* http://unfccc.int/resource/docs/2002/sb/inf02.pdf

(GHG) emissions they will need to implement domestic climate policies, or make use of the offshore Kyoto mechanisms (see below).

Of the large OECD countries, Germany and the UK achieved significant reductions of 19 per cent and 13 per cent respectively. However, Germany's apparently green performance is in fact a statistical blip, caused by reunification, and not the result of implemented climate actions. In the UK much, if not all, of the reduction in greenhouse gas emissions has happened because of reasons other than implemented climate actions (such as the 'dash for gas' and long-term structural economic change away from energy intensive manufacturing industry and towards services). The UK government has been unable to supply hard evidence that this is not the case (the effectiveness of mitigation policies are hard to prove!). Germany and the UK have thus benefited from one-off climate bonuses. Both countries are unlikely to be so lucky in meeting their Kyoto targets in the years to come.

Developing countries have been highly critical of the failure to date of the industrialized countries to meet their original 1992 promise of stabilizing emissions at 1990 levels by the year 2000. The G77 and China group point out that the lack of progress at national levels in establishing meaningful and effective climate policies and programmes means *in practice* governments are not applying the precautionary principle – even the weak version (see Figure 5.25).

**Figure 5.25** How certain does the science have to be before developed nations will take the lead? A question the G77 and China group want the developed nations to answer.

While developed countries will face climate impacts, particularly for high increases in global temperature, their short-term political concerns tend to focus on the economic, social and political pain involved in weaning their economies off fossil fuels and towards alternative energy technologies.

Reducing the cost of compliance with Kyoto has been of paramount importance to the big players from the developed world, the USA and Europe in particular. These countries lobbied hard to have offshore Kyoto mechanisms (sometime referred to as 'flexibility mechanisms'). As you may imagine, these mechanisms are themselves a source of political conflict. The latest economic assessments reviewed in the IPCC TAR suggest that compliance costs are halved through emissions trading among Annex I countries and would be halved again if global trading (that is, with developing countries) were allowed.

## 5.5  Conflicts over the Kyoto mechanisms

The emissions trading mechanism and the clean development mechanism (CDM) are described in Box 5.6. They were introduced as a way of reducing the economic costs of meeting climate targets – a sort of economic 'pain killer' if you like. If, at some point, it is cheaper for Canada, for example, to reduce someone else's greenhouse gas emissions rather than its own, then it is possible for Canada to buy those reductions and claim them as if they were achieved at home.

emissions trading
mechanism

### Box 5.6  Emissions trading and clean development mechanisms

**Emissions trading mechanism**

International emissions trading allows industrialized countries to balance their greenhouse gas books by buying and selling greenhouse gas pollution permits (see Chapter Three on tradeable permits). If Canada (for example) is struggling to meet its Kyoto targets and cannot find emission reductions as cheaply at home as they are at internationally traded prices, then it can purchase from another country (say Russia) a certain amount of that country's emission entitlements. This is providing that Russia is in a position where its actual emissions are less than its target and it therefore has emission entitlements to sell. Emission entitlements (certified by permits) are then deducted from the recorded inventory for Russia and added to the recorded inventory for Canada, in a simple double-entry book-keeping manner. For example if Canada has purchased 10 million tonnes (mt) $CO_2$ emission entitlements from Russia, then Canada is deemed to have emitted 10 mt $CO_2$ less than the volume recorded and Russia 10 mt $CO_2$ more. In effect Canada can emit 10 mt $CO_2$ more than its original target, but Russia must emit 10 mt $CO_2$ less than its original target (Figure 5.26)

The Kyoto Protocol does not just cover carbon dioxide ($CO_2$). It also covers emissions of methane ($CH_4$), nitrous oxide ($N_2O$), hydrofluorocarbons (HFCs), perfluorocarbons (PFCs) and sulphur hexafluoride ($SF_6$).

Emissions trading will require a careful system of greenhouse gas inventory book-keeping for the basket of six greenhouse gases from all sources. National, regional or multinational emissions trading schemes can be far more complex and can involve the simultaneous trade in carbon, renewable and green energy certificates. The Kyoto Protocol, however, deals only with the simple double entry book-keeping version between states.

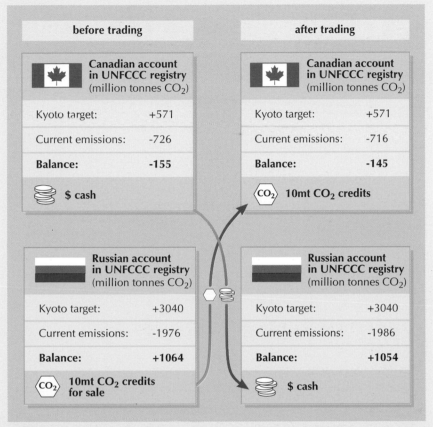

Figure 5.26 Schematic representation of emissions trading between, for example, Canada and Russia: a negative balance means a state is in debit – its emissions are greater than its target, and vice versa.

## Clean development mechanism (CDM)

clean development mechanism

The CDM is about balancing the greenhouse gas books by counting emissions reductions achieved through overseas projects specifically designed to reduce emissions and support sustainable development. Let us suppose Canada sees an opportunity to help China reduce its greenhouse gas emissions by 10 mt $CO_2$ by choosing a cleaner technology or by changing some practice or system. The two countries agree to cooperate to reduce the greenhouse gases from Chinese coal plants, for example, by upgrading the technology or changing some other aspect of its system. China gets the benefits of better technology (hopefully with all the attendant secondary economic benefits in terms of technology transfer, reduced local air pollution and cheaper energy) while Canada claims the credit for the greenhouse gas emissions avoided as a result of using the better technology. Everyone is happy – at least in theory (Figure 5.27).

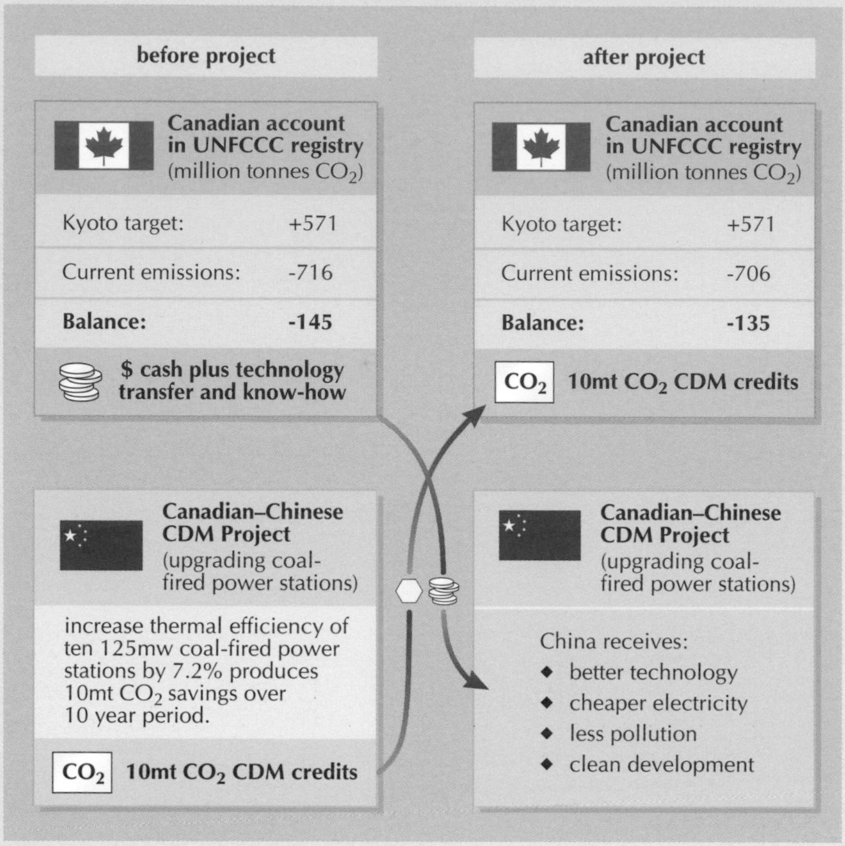

| before project | after project |
| --- | --- |
| **Canadian account in UNFCCC registry** (million tonnes $CO_2$) | **Canadian account in UNFCCC registry** (million tonnes $CO_2$) |
| Kyoto target: +571 | Kyoto target: +571 |
| Current emissions: -716 | Current emissions: -706 |
| **Balance:** **-145** | **Balance:** **-135** |
| **$ cash plus technology transfer and know-how** | $CO_2$  10mt $CO_2$ CDM credits |
| **Canadian–Chinese CDM Project** (upgrading coal-fired power stations) | **Canadian–Chinese CDM Project** (upgrading coal-fired power stations) |
| increase thermal efficiency of ten 125mw coal-fired power stations by 7.2% produces 10mt $CO_2$ savings over 10 year period. | China receives: <br> • better technology <br> • cheaper electricity <br> • less pollution <br> • clean development |
| $CO_2$  10mt $CO_2$ CDM credits | |

**Figure 5.27** Schematic representation of clean development mechanism; in this example Canada and China cooperate.

However the CDM is not without its critics. One key criticism of the CDM is that it may in some cases act as a perverse incentive by actually increasing greenhouse gas emissions relative to what they might otherwise have been (the so-called project 'baseline'). This could possibly arise if a developing country delayed installing or upgrading to cleaner technology in the hope that it could eventually find a CDM partner to share these costs. Other criticisms of the CDM include the following: it involves high administrative costs; it will be difficult to police; it favours larger multinational firms; and it is open to abuse by either party ('donor' countries and developing 'host' countries).

Some critics of the two Kyoto mechanisms argue that the mechanisms are simply a way for developed countries to avoid reducing their own emissions 'at home'. Internal trading within the Annex I block – between Western European countries and EIT countries – does not affect the Kyoto commitment for the industrial group of countries as a whole. The Annex I 'bubble' must end up 5 per cent lower than 1990 by 2012. If used to any significant degree, the CDM represents a long-term source of cheap emission permits for developed countries. Unfettered use of the mechanism would allow rich countries to avoid demonstrating leadership on reducing greenhouse gas emissions at home for years to come. For these reasons, the small print of the rules of the Kyoto Protocol dictates that the developed countries must meet their targets mainly through domestic measures and not by relying on the use of emissions trading or the CDM. However, like much of the small print in MEAs, the rule is vague and open to considerable political interpretation.

Under the rules of the Kyoto Protocol, countries are only permitted to trade emissions or engage in the CDM if they meet certain eligibility criteria. Also in the case of the CDM, only certain cleaner technologies are eligible including renewable energy technologies and energy efficiency projects and programmes. The introduction or promotion of new energy technologies is capable of generating considerable public debate (**Blowers and Elliott, 2003**). Projects involving nuclear power are not eligible, to the dismay of the nuclear industry that campaigned hard to project nuclear power as a viable part of the solution to climate change (Figure 5.28).

**Figure 5.28** The nuclear industry has gone to great lengths to persuade the climate negotiations that nuclear power is a key part of the solution to the issue of climate change.

Under the emerging climate regime, emission reductions achieved as a result of net uptake of carbon from the atmosphere from land use change and forestry activities are also counted against national emission reduction targets and as part of the CDM. Complex rules and procedures for the correct treatment and eligibility of agricultural or forest projects are emerging. A good deal of political

You can see an illustration of monocultural forest plantation in Chapter Six, Figure 6.3b.

attention has been given to the issue of forests as carbon 'sinks' in the negotiations. The environmental benefits of planting large monocultural stands of trees to offset carbon emissions elsewhere in the economy are contested by various groups. Environmental NGOs have raised ecological concerns about monocultural plantations which have a lower biodensity than natural forest. In some cases, monocultural plantations can prevent natural adaptations to climate change rather than assisting them by acting as a 'corridor' (as natural forests do). The impact of plantations on soil quality is fiercely contested between environmentalists and forestry companies; plantations can significantly alter hydrological systems compared with a forest or other natural ecosystem; plantations can trigger other processes to release carbon dioxide and other greenhouse gases which in some circumstances may cancel out any benefits from the carbon sequestered in the growing plantation. Another argument is that forests are only sinks for $CO_2$ while they are growing and in this sense represent a small, one-off measure to delay real mitigation (such as switching to renewable energy sources and energy efficiency).

Technologically, the existence of the UNFCCC and the climate negotiation process has given a large boost to climate-friendly technologies and industries, while at the same time posing a significant business threat to the fossil fuel industries. However, the jury is still out on the extent to which economic approaches at the global level will be effective; this depends largely on the fate of the Kyoto Protocol and its emissions trading and CDM mechanisms. Even at national levels there is still little compelling evidence of the overall effectiveness of economic approaches to greenhouse gas emission reductions – governments have yet to pull the policy levers sufficiently to get anywhere close to emissions cuts required to meet the objective of the convention.

## 5.6 The interests of the vulnerable and the poor

Principle 2 of the UNFCCC (Box 5.3), highlights the 'specific needs and special circumstances of developing country Parties, especially those that are particularly vulnerable to the adverse effects of climate change'. This is a deliberately vague, but important phrase. It is shorthand for the developing country concerns about the impacts of climate change and the implications for equity and sustainable development.

Some of the most vulnerable countries are also some of the world's least developed (Figure 5.29). Many small island developing states in the South Pacific as well as low-lying coastal countries such as Bangladesh, for example, face the significant threat of half a metre or so sea level rise. African countries such as Ethiopia, Mozambique and Zambia are extremely worried about the possibility of more frequent and more severe droughts. All developing countries, but particularly those already sensitive to climate variability, are concerned about the possible impacts of climate change. Many of them have received small amounts of bilateral and multilateral financial assistance to help them assess their vulnerability to climate change (typically in the order of US$ 100,000 per country). But however much money is spent on assessments, future uncertainties about the nature and extent of climate change mean that it is difficult for developing countries to decide on their priorities in terms of requests for assistance or compensation under the UNFCCC. Some countries, for various reasons, have not even conducted adequate vulnerability assessments.

Developing countries frequently request assistance from donor countries (the developed nations) and the development banks to help develop their capacities to adapt to actual or predicted climate change in their region. It will be some time yet before funding for real adaptation projects (e.g. sea walls, new drought and saline resistant crops, dams etc) gets properly underway. This is a major source of frustration for developing countries and greatly affects the politics of climate

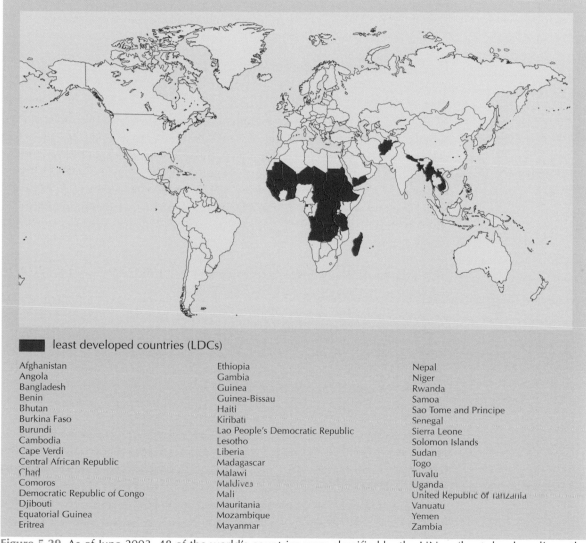

**least developed countries (LDCs)**

| | | |
|---|---|---|
| Afghanistan | Ethiopia | Nepal |
| Angola | Gambia | Niger |
| Bangladesh | Guinea | Rwanda |
| Benin | Guinea-Bissau | Samoa |
| Bhutan | Haiti | Sao Tome and Principe |
| Burkina Faso | Kiribati | Senegal |
| Burundi | Lao People's Democratic Republic | Sierra Leone |
| Cambodia | Lesotho | Solomon Islands |
| Cape Verdi | Liberia | Sudan |
| Central African Republic | Madagascar | Togo |
| Chad | Malawi | Tuvalu |
| Comoros | Maldives | Uganda |
| Democratic Republic of Congo | Mali | United Republic of Tanzania |
| Djibouti | Mauritania | Vanuatu |
| Equatorial Guinea | Mozambique | Yemen |
| Eritrea | Mayanmar | Zambia |

**Figure 5.29** As of June 2003, 48 of the world's countries were classified by the UN as 'least developed', on the basis of average per capita incomes of less than US$2.47 per day. *Source:* UN, 2003.

change. Developing countries want to eradicate poverty and develop economically. To do this they emphasize their need for advanced technologies and for building appropriate capacity (training, education, institutional strengthening). However, developed countries (in particular the USA and Europe) emphasize the need for developing countries to create the right sort of 'enabling conditions' for their economies to grow. Some suggest that this is simply a way for the powerful OECD trading countries to try to gain access to developing country markets. Precisely how developed countries should help to build the developing countries' economies is a major source of political argument.

In theory, there is consensus that the first priority of developing countries is poverty eradication and economic development. The problem comes in interpreting what *kind* of economic development. All sides emphasize the need for 'sustainable economic development', but beyond the rhetoric there is no shared vision of what exactly *sustainable* economic development means.

## 5.7  The interests of fossil fuel producing and exporting nations

In the absence of significant technological breakthroughs, if the UNFCCC actually achieved its ultimate goal, it would almost certainly mean the collapse of the global trade in most fossil fuels. The prospect of losing vital oil revenues has tended to focus the minds of climate negotiators from OPEC (Organization of Petroleum Exporting Countries) who wield considerable power in the negotiations. They consider the potential decline of the global petroleum industry as a 'climate impact' just as critical to them as sea level rise, for example, is to small island states.

Here is an example of a typical intervention in negotiations from Saudi Arabia on the matter:

> We expect the emergence of solid recommendations of the kind of action required to minimize the impacts on Non-Annex I country Parties. They may include.... increase of foreign investment in the affected Non-Annex-I Parties to assist them to diversify their economies and reduce their heavy dependence on the exportation of fossil fuels.

(UNFCCC, 1999)

In other words, OPEC expects compensation for any loss in oil revenues that they may face as a result of decisions taken on climate change. In one sense, OPEC is just an extreme example of the political sensitivities that exist in most developed countries about the risks inherent in our response to climate change. With the exception of the US Government under the Presidency of George W. Bush, few governments in developed countries have been honest and open about their deep-rooted concerns that Kyoto ultimately represents a significant threat to current patterns of production and consumption and therefore economic and political structures.

# 5.8 Conflicts around the principle of 'equity'

Principle 1 of Article 3 (Box 5.3) states that 'Parties should protect the climate system for the benefit of present and future generations of humankind, on the basis of *equity* and in accordance with their common but differentiated responsibilities and respective capabilities.' What represents an equitable approach to sharing the climate?

## Activity 5.5

Carbon dioxide emissions from the Annex I and Non-Annex I regions in 2000 stood at 3.7 and 2.7 GtC per year respectively (excluding most emissions from ships and aircraft). In the period 1980–2000, emissions from the Non-Annex I region doubled. Assuming Annex I emissions remained at 3.7 GtC per year and that Non-Annex I emissions continue to grow in the next 20 years business as usual (i.e. they double again by 2020 to 5.4 GtC per year), when would developing countries overtake developed countries as the main source of CO2 emissions?

You can find the answer at the end of the chapter.

Scientifically, each and every tonne of carbon (or other greenhouse gas) reduced, from whatever technology or process and from wherever in the world, is equally important. Within the next decade or so, developing countries will overtake the developed world as the largest emitters of greenhouse gases. Since the legally binding emissions reduction targets in the Kyoto Protocol apply only to developed countries, by 2010 the treaty will effectively only cover half of the climate change problem. That fraction will then continue to decline further into the future. Therein lies a major political stumbling block in the negotiations. Developed countries quite reasonably point out that developing countries also need to honour their commitments in the UNFCCC (Box 5.4) by reducing their future greenhouse gas emissions. Countries like India and China recognize the scientific basis of this perspective, but argue that they have not yet had the same chance to develop economically using fossil fuels as the OECD nations have had throughout the Industrial Revolution.

Developing countries such as India and China find themselves on the horns of a dilemma. They too will be victims of climate impacts and do not have the resources that developed countries do to cope with the consequences. Reducing greenhouse gas emissions globally is certainly in their interest. However, they are also eagerly hoping, planning and acting for rapid economic development (Figure 5.30). In turn this requires lots of cheap energy. Any global agreement that makes energy more expensive is an unwelcome brake on the economic aspirations on any developing economy.

For any hope of achieving the ultimate goal of the UNFCCC, global emissions have to decrease by as much as 60–90 per cent during the course of this century according to climate models. This means developing countries are going to have to slow down, stabilize and even reduce their emissions *before* they have had a chance to develop. Rich industrialized countries cannot solve the problem of climate change on their own. This partly explains the context of President George W. Bush's well-publicized and contoversial (Figure 5.31) rejection of the Kyoto Protocol in 2001: he believed Kyoto is unfair because it lets developing countries off the hook.

So what would be an equitable approach to sharing out the limited 'emissions space' humans would have left in which to stabilize the climate? All 187 Parties to the UNFCCC agree they are in the same boat and are therefore all, in some ways, responsible for stabilizing concentrations. They share the same atmosphere and will all, on balance, suffer in some way or other from rising temperatures and their various physical and socio-economic effects. Given the scientific uncertainty, it is difficult to specify what level of $CO_2$ increase is 'safe'. We cannot really say how safe 'safe' is. But clearly the sooner we can stabilize emissions and thereby lower atmospheric concentrations, the lower the likely risks we, or rather our grand-children, will face in the future. The EU has suggested a long-term climate target of limiting the maximum global temperature increase to 2 °C. However, this is the closest example to date of an official political attempt to quantify climate risk.

There is at present no shared agreement on what constitutes an equitably differentiated approach to allocating future climate commitments, nor is there any prospect of one emerging for some time. But if climate change continues to worry us and if there is political will to act, then at some stage the thorny question of who is prepared to give up which 'rights' will have to be addressed. On the one hand there are rights to a safe atmosphere, on the other there are rights to *sustainable* economic development.

Table 5.2 lists the carbon dioxide emission levels in tonnes per capita ($tCO_2$/capita) in 2000 for various countries. As you can see from the table there is a large range of values for the emissions per capita. The average UK citizen generates 4 times the emissions of the average Chinese citizen, about 10 times the average Indian citizen and more than 180 times a Tanzanian citizen. These gaps would double again if we were to make the comparison with the USA.

**Figure 5.30** The Chinese economy is booming: under-pinning Beijing's growth and that of the other major cities is an insatiable demand for cheap fossil fuels.

**Figure 5.31** US students protesting against their leader's rejection of the Kyoto Protocol outside UNFCCC COP 6, Bonn, Germany, July 2001.

**Table 5.2**   Per capita $CO_2$ emissions ($tCO_2$) for selected countries, 2000

| Country | $CO_2$ emission levels per capita ($tCO_2$/capita) |
|---|---|
| China | 2.4 |
| Honduras | 0.7 |
| India | 0.9 |
| Nicaragua | 0.7 |
| Pakistan | 0.7 |
| United Kingdom | 8.9 |
| USA | 20.6 |
| United Republic of Tanzania | 0.05 (smallest per capita polluter) |
| Qatar | 60.0 (largest per capita polluter) |
| world average | 3.9 |

*Source:* IEA, http://www.iea.org/statist/keyworld2002/key2002/keystats.htm

One suggestion about a fair approach to managing the earth's climate is based on the notion that the atmosphere is a global commons and everyone should have equal rights to it. The idea is to contract global emissions down to safe levels and then allocate equal rights to pollute the earth's atmosphere to every person on the planet. Of course the world is a very uneven and inequitable place today, so there would have to be a transition period of convergence towards this fairer climate world. The idea has been labelled **'contraction and convergence' (C&C)**.

contraction and convergence (C&C)

### Activity 5.6

The key idea behind the concept of C&C is that to stabilize greenhouse gas concentrations, (a) global emissions need to contract significantly down to some safe threshold level and (b) that the threshold level of pollution be shared equitably on a per capita basis among all nations. The following simple calculations will illustrate the concept further.

Global population is projected to reach 9 billion in 2050. Imagine that the C&C approach was accepted by the UNFCCC and that it set a global greenhouse gas emissions stabilization target in 2050 at 3 GtC per year, half the rate compared with 2000 (6.4 GtC per year). If every person on the planet in 2050 had an equal right to pollute the atmosphere, what individual yearly emissions budget would each person be allocated?

(1 Gt is 1 billion tonnes.)

For the answer, go to the end of the chapter.

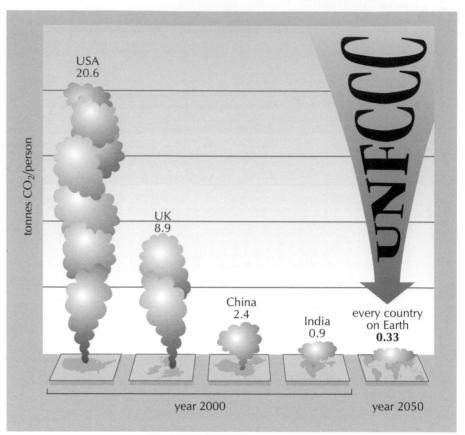

**Figure 5.32** An illustration of the implications of contraction and convergence if it were adopted by the UNFCCC process. In this example, global emissions contract by 50 per cent by 2050 and are then shared equally among 9 billion people.

The destination for a future climate regime based on the application of the contraction and convergence approach to climate equity is significantly different from the situation today (Figure 5.32).

There is no doubt that the politics of climate change will continue to tiptoe around the notion of equity. Those who currently face the worst impacts of climate change are frequently the poorest nations and therefore those with the least ability to cope with the additional problems this might bring, such as drought, floods and disease. The poorest developing countries will continue to look towards rich OECD neighbours for compensation and help in adapting to climate change. They will also be demanding leadership and action. Large and powerful developing nations such as China and India are caught in the middle. They are just a few decades away from economies considerably closer to those of Europe and North America than some of their poorer developing country neighbours.

The way out of this dilemma would seem to be for industrialized nations to accelerate their climate action plans to reduce greenhouse gas emissions at home and for leading developing countries to start to talk about the actions they would be prepared to take to reduce their greenhouse gas emissions and on what timescale.

# Summary

- The UNFCCC is an example of a soft MEA. As its name suggests, it is a framework consisting of an overriding objective, principles and commitments.

- The UNFCCC is a living process of engagement and dialogue between nations.

- The Kyoto Protocol is an example of a hard MEA. It is legally binding and has an elaborate compliance and enforcement mechanism. Under the Protocol, greenhouse gas emissions from developed countries will be gradually and successively reduced in order to meet the objective of the UNFCCC. The first group target for developed countries is to reduce their emissions by 5 per cent by 2012 compared with 1990. Countries must meet their individual Kyoto targets mainly through domestic measures but also through two Kyoto mechanisms: international emissions trading and the clean development mechanism.

- Political uncertainty and conflict is inherent in the design of the UNFCCC and its Kyoto Protocol. These include disputes about the principle of leadership, conflicts about the appropriate use of emissions trading and the CDM, the specific interests of the vulnerable and least developed nations, the concerns of the oil exporting economies, and the interests of powerful developed and developing governments in maintaining, or aspiring to, current patterns of production and consumption.

- The atmosphere is a global commons. The interpretation of the principle of equity as the allocation of equal per capita emissions rights for every person on Earth gives a practical quantification of sustainable development. The answer is of the order of $0.33 - 1$ tonne of $CO_2$ per person per year (Figure 5.32 is illustrative of the rough order of magnitude), one tenth of the per capita level in the UK in 2000 and one twentieth of that in the USA in 2000.

# 6   Conclusion

Climate change calls for a precautionary approach to environmental decision making under high levels of scientific and political uncertainty. Indeed, as we have seen, uncertainty in one feeds the other. However, uncertainty has not prevented us from taking significant precautionary action. The international response is a living negotiation fed by the best available scientific evidence. The UNFCCC and its Kyoto Protocol are the first steps in the international response to climate change. In the years to come we will get the chance to judge how effective they are as tools to manage environmental change at the global level. We are at the start of a path of learning by doing. Future agreements will take into account evolving scientific evidence as well as political agreements/disagreements as they unfold on a sequential basis.

Political conflicts about the most appropriate way to respond to climate change have their roots in several areas including: the scientific uncertainties inherent in

**Figure 5.33** A UNEP cartoon: sustainable development is a delicate balance in pursuit of the three priorities – economy, environment and society.

our understanding of the climate system and its future behaviour; uncertainties about the effectiveness of the 'wait and see' approach versus intervention in the form of adaptation and mitigation response strategies; and different perspectives on key principles of soft international law on sustainable development, notably the precautionary principle, the principle of common but differentiated responsibility (leadership) and the principle of equity.

Very few, if any, governments spend their time thinking mainly about climate or indeed environmental issues in general. On the whole, all governments are united in their pursuit of economic and social development as primary objectives. When climate change has to take its place in a competing list of government objectives, conflicts arise. Climate impacts are part of a larger question of how complex systems interact and shape prospects for sustainable development.

Hence, we are in a phase in the evolution of our global political approach to climate change that can be characterized as 'looking but not touching'. For all the talk about the threat of climate change, we have yet to witness international action on the scale necessary to have any chance of stabilizing greenhouse gas concentrations. Politicians and technocrats are busy looking at the costs and benefits associated with both the climate adaptation and mitigation response options. They are searching for 'no regret' win–win sustainable development policy solutions: the right balance of tools that will lead to a safe climate, and a prosperous global economy and vibrant human society (Figure 5.33).

This may not prove possible. At some stage, they may have to grasp the problem of the costs of dealing with climate change. Buckle your seatbelt. We are at the start of a long-term sequential decision-making process under high levels of scientific and political uncertainty.

# References

Bingham, N. and Blackmore, R. (2003) 'What to do: how risk and uncertainty affect environmental issues' in Hinchliffe, S.J. et al. (eds).

Bingham, N., Blowers, A.T. and Belshaw, C.D. (2003) *Contested Environments*, Chichester, John Wiley & Sons/The Open University (Book 3 in this series).

Blackmore, R. and Barratt, R.S. (2003) 'Dynamic atmosphere: changing climate and air quality' in Morris, R.M. et al. (eds).

Blowers, A.T. and Elliott, D.A. (2003) 'Power in the land: conflicts over energy and the environment' in Bingham, N. et al. (eds).

Blowers, A.T. and Smith, S.G. (2003) 'Introducing environmental issues: the environment of an estuary' in Hinchliffe, S.J. et al. (eds).

Burgess, J. (2003) 'Environmental values in environmental decision making' in Bingham, N. et al. (eds).

Hinchliffe, S.J., Blowers, A.T. and Freeland, J.R. (2003) *'Understanding Environmental Issues'*, Chichester, John Wiley & Sons/The Open University (Book 1 in this series).

Hurrel, A. and Kingsbury, B. (1992) 'International environmental law: its adequacy for present and future needs' in Hurrel, A. and Kingsbury, B. (eds) *The International Politics of the Environment*, Oxford, Clarendon Press.

IPCC (2001a) 'Climate change 2001: The scientific basis', *Contribution of Working Group I to the Third Assessment Report of the Intergovernmental Panel on Climate Change*, Cambridge, Cambridge University Press.

IPCC (2001b) 'Climate change 2001: impacts, adaptation and vulnerability', *Contribution of Working Group II to the Third Assessment Report of the Intergovernmental Panel on Climate Change*, Cambridge, Cambridge University Press.

IPCC (2001c) 'Climate change 2001: mitigation', *Contribution of Working Group III to the Third Assessment Report of the Intergovernmental Panel on Climate Change*, Cambridge, Cambridge University Press.

Morris, R.M., Freeland, J.R., Hinchliffe, S.J., and Smith, S.G. (2003) *Changing Environments*, Chichester, John Wiley & Sons/The Open University (Book 2 in this series).

Reddish, A. (2003) 'Dynamic Earth: Human Impacts' in Morris, R.M. et al. (eds).

Taylor, A. (2003) 'Trading with the environment' in Bingham, N. et al. (eds).

UN (2003) http://www.un.org/special-rep/ohrlls/ldc/list.htm (accessed 23 June 2003)

UNFCCC (1999) FCCC/SB/1999/MISC.6; http://cop5.unfccc.int/resource/docs/1999/sb/misc06.htm

# Answers to Activities

## Activity 5.2

Statement A could be associated with number 3. You could also associate it with number 1.
Statement B could be associated with number 1.
Statement C could be associated with number 2.
Statement D could be associated with number 3.
Statement E could be associated with number 4.

## Activity 5.3

Article 2 is a strong interpretation of the precautionary principle in that it seeks to avoid dangerous anthropogenic interference with the climate system. However the sentence 'such a level should be achieved within a time-frame sufficient to ... enable economic development to proceed in a sustainable manner' clearly opens up the possibility of a weak interpretation of precaution, i.e. let's act, but only if it does not also interfere with *sustainable* economic development.

## Activity 5.4

Approx 24%. If total emissions in 1990 were say 100 units, then without Kyoto they would have naturally (on a business as usual basis) grown by 22 x 0.8% or 17.6%. If Kyoto is achieved they will actually stand at 95 units in 2012. So the headline 5% Kyoto target may represent a real reduction in emissions from Annex 1 countries of approximately 24% [((117.6/95)-1) x 100%)] compared to the business as usual case.

## Activity 5.5

2007 approximately. A doubling of Non-Annex I emissions from 2.7 to 5.4 GtC per year is equivalent to an annual increase of 0.135 GtC per year (2.7/20). For Non-Annex I emissions to grow by 1GtC per year from 2.7 to 3.7 GtC per year, it would take just over 7 years (1/0.135 = 7.4 years). More accurately, the rate of growth of Non-Annex 1 emissions was nearly 3% in the period 1990–2000 compared with nearly 6% in the period 1980–2000. In reality, therefore, it may be about 2012–2015 before developing countries overtake the developed as the largest emitters of greenhouse gases.

## Activity 5.6

0.3 $tCO_2$ per person per year. 3 gigatonnes is $3\times10^9$ tonnes or 3 billion tonnes. So we have 9 billion people allowed to emit 3 billion tonnes. Each person is therefore allowed to emit 3/9 tonnes carbon dioxide each year. This is 0.33 tonnes per capita per year. This level of per capita emissions corresponds to roughly half the 1999 per capita emissions in countries such as Nicaragua, Honduras and Pakistan.

# Chapter Six

# Sustainable environmental futures: opportunities and constraints

Steve Hinchliffe, Andrew Blowers and Joanna Freeland

# Contents

# 1    Introduction

How can the needs of current generations be met without compromising the life chances of future generations? This is the basic question that those advocating sustainable development have set for politicians, businesses and the collective human population (see **Blowers and Smith, 2003**). In a world where currently 80 million people are in danger of starvation and ten times that figure (or nearly one sixth of the total human population) live in conditions of poverty, the challenge of sustainable development is acute. Not only do we need to make sure that we safeguard the future, we also need to make sure that those living on the planet right now are given a fair chance of a decent life. To achieve this it is necessary to recognise that human wellbeing and the wellbeing of a whole host of other living beings are interlinked. Similarly, the wellbeing of living creatures is dependent upon the functioning of many non-living components of environments. Human life chances now and in the future will be reduced if the environments in which people co-habit, with other people and with non-humans, are systematically degraded. Similarly, the chances are that unless we do something about the current inequalities of the social world we will have very little chance of acting in ways that are less destructive of the planet and its life support systems.

In this chapter we address the question 'What are the constraints and opportunities for sustainable development?' We do so largely through the example of biodiversity, although we will also touch on other areas of environmental concern. By choosing to focus on biodiversity at the end of this book we do not wish to suggest that this is somehow a privileged topic. However, it does have the advantage of returning us to the issue of extinction, which loomed large in the first book in this series (**Hinchliffe et al., 2003**).

The rest of this chapter is divided into five sections. Section 2 introduces you to biodiversity, looking at the term's relatively recent invention and examining whether or not its rise to prominence offers a significant opportunity for sustainable environmental futures. The third section looks at economic attempts to stem the current trend of biodiversity losses by re-valuing biodiversity as an economically attractive asset. This section asks whether or not it is possible to conserve biodiversity at the same time as securing economic development. Is it possible, in other words, to have your cake and eat it? This leads us, in Section 4, to a discussion of ecological modernization, a set of approaches to social development that seems to offer a means of securing economic livelihoods at the same time as sustainable environments.

If Sections 2 to 4 allow us some optimism in that we are managing to work out ways of valuing human and non-human wellbeing and squaring the circle of environment and development, Section 5 is more cautious. There we look at the enduring political issues that constrain our attempts to find ways of living sustainably on the planet. Notably, we highlight the role of global inequalities as a major constraint on sustainable development. Finally, in the last section, we ask whether or not, in the light of the arguments and examples presented in the chapter, we can be optimistic regarding future environments. We look at some local-scale attempts to foster biodiversity and ask whether or not they can add up to a more significant environmental movement.

Five aims should be borne in mind as you work through the chapter:

- to identify constraints and opportunities for sustainable development;
- to identify the importance of inequalities as a constraint on sustainable development;
- to describe scientific, economic and political risks and uncertainties with respect to the future conservation of biodiversity;
- to judge the extent to which ecological modernization can deliver sustainable futures;
- to set up optimistic and pessimistic arguments regarding environmental futures.

# 2   Biodiversity – the concept as a response

Humans have been interested for thousands of years in biological diversity, or the range of organisms with which we co-exist. For example, throughout history and in many parts of the world people built up detailed informal understandings of plant locations and properties, and used these understandings to develop foods and medicines. In the 18th and 19th centuries, natural historians collected specimens from all over the world and painstakingly named and classified samples that can today be found in many museum collections. Biological diversity led Gregor Mendel to establish the principles of genetics and Charles Darwin to propose the theory of evolution by natural selection (Figure 6.1).

**Figure 6.1** Moritz Rugenda's engraving *La Forêt du Brésil*. This engraving fired Darwin to visit the Tropics, and he later wrote that the real rainforest was even more luxuriant.

So biological diversity has had, and continues to have, instrumental value for people (see **Hinchliffe and Belshaw, 2003** for a consideration of this term). But there is another side to this human interest. Consider the following quotation written more than fifty years ago (when it was still considered acceptable to use gender-specific language) by Aldo Leopold, widely acknowledged as one of the pioneers of wildlife conservation in North America:

> Our grandfathers were less well-housed, well-fed, well-clothed than we are. The strivings by which they bettered their lot are also those which deprived us of [passenger] pigeons. Perhaps we now grieve because we are not sure, in our hearts, that we have gained by the exchange. The gadgets of industry bring us more comforts than the pigeons did, but do they add as much to the glory of the spring?
>
> It is a century now since Darwin gave us the first glimpse of the origin of the species. We know now what was unknown to all the preceding caravan of generations: that men are only fellow-voyagers with other creatures in the odyssey of evolution. This new knowledge should have given us, by this time, a sense of kinship with fellow-creatures; a wish to live and let live; a sense of wonder over the magnitude and duration of the biotic enterprise.
>
> Above all we should, in the century since Darwin, have come to know that man, while captain of the adventuring ship, is hardly the sole object of its quest, and that his prior assumptions to this effect arose from the simple necessity of whistling in the dark.
>
> These things, I say, should have come to us. I fear they have not come to many.
>
> For one species to mourn the death of another is a new thing under the sun. The Cro-Magnon who slew the last mammoth thought only of steaks. The sportsman who shot the last [passenger] pigeon thought only of his prowess. The sailor who clubbed the last auk thought of nothing at all. But we, who have lost our pigeons, mourn the loss. Had the funeral been ours, the pigeons would hardly have mourned us. In this fact, rather than in Mr. DuPont's nylons or Mr. Vannevar Bush's bombs, lies objective evidence of our superiority over the beasts.
>
> (Leopold, 1948, pp.109–10)

The irony implicit in Leopold's writing is the apparently unique ability of humans consciously to mourn the demise of other species, while at the same time single-handedly inflicting more harm upon other organisms than has probably been done by any other species in history.

The more attention that people paid to species and habitats, the more they began to realize that biological diversity was not something that should be taken for granted. In 1985, the Society for Conservation Biology, which sought to promote biological diversity and publicize the plight of many species and habitats, was founded in the USA. And in the mid-1980s the term biodiversity was introduced as a contraction of 'biological' and 'diversity'. The term received its first large-scale international airing in New York in 1986, at the *National Forum on BioDiversity*. This meeting brought together many biologists, plus some representatives from the social sciences and arts. The main focus was the current state of global biodiversity, with particular attention paid to ongoing threats, and ways in which biodiversity could be preserved. The major concerns voiced by participants were the increasingly accelerated rates of species extinction associated with the ongoing destruction of habitats around the world. Tropical forests were dealt with extensively because they are highly diverse areas that are being rapidly destroyed. At the same time, attention was paid to a diversity of other ecosystems including temperate forests, grasslands, coastal zones, salt marshes and oceanic islands – all united by a common theme of habitat loss and the decline of species. The proceedings of this conference are published in *Biodiversity*, a text that was edited by E.O. Wilson, one of the conference organizers and a zoologist of international standing who has spent much of his life studying biodiversity (Wilson and Peter, 1988).

## 2.1  What is biodiversity?

Was the coining of this new word 'biodiversity' a significant intellectual and political response in the fight to protect life on earth? Before we answer this question we will look at a more detailed description of what biodiversity actually means. In the following discussion, bear in mind that the concept of biodiversity is often linked to that of conservation biology. This is in part because assessments of biodiversity frequently alert people to the decline of diversity, and this in turn leads to calls for conservation policies aimed at reversing or at least slowing the loss of species and habitats within particular areas.

There are three elements to the description of biodiversity. First, **species diversity** refers to the number of species present in a particular time or place. Second, **ecosystem** or **habitat diversity** refers to the variety of ecosystems and/ or habitats in an area. Third, **genetic diversity** refers to the variety of genetic material within a species or population. We can look at each in turn.

species diversity

ecosystem or habitat diversity

genetic diversity

### Species Diversity

Historically, counting species has been the most common way to measure diversity – for example, estimates of how extinction rates have changed over time rely on comparisons of the numbers of species that have gone extinct. Species counts may be enhanced by estimating the numbers of individuals within each species, and are by no means straightforward (see the discussion in Section 2.2 and **Freeland, 2003**).

## Ecosystem Diversity

The primary focus of biodiversity has more recently been moving away from species to the ecosystems in which they live (see **Morris and Turner, 2003**). In terms of conservation, it is generally more desirable to preserve ecosystems or habitats as opposed to targeting single species (whose habitat may then be lost, making species preservation unsustainable). As people become more aware of the link between species loss and habitat loss, they can appreciate that the preservation of an ecosystem or habitat involves the preservation of countless species. At a global level, the diversity of habitats and ecosystems is under threat. Forests are home to much of the known terrestrial biodiversity, but are being depleted in many areas (see **Morris, 2003a**). Despite regrowth in some places, many forests are still shrinking rapidly, particularly in the tropics and in northern Russia. A third of the world's coral reefs – among the richest ecosystems – face collapse (see **Blackmore and Barratt, 2003**). Likewise, coastal mangroves, a vital habitat for countless species, are also vulnerable to pollution, sea-level fluctuations and climate change.

## Genetic Diversity

The third measure of biodiversity, genetic diversity, is also playing an increasingly important role in conservation biology and assessments of diversity. Technological advances mean that it is now relatively easy to quantify the genetic diversity of individuals, populations and species. For example, we now know that the number of genes for each individual is about 1,000 in bacteria, about 10,000 in some fungi and 400,000 or more in many flowering plants. A typical mammal such as the house mouse (*Mus musculus*) has about 100,000 genes. Although there is considerable similarity between the genes that are shared by individuals within a species such as the house mouse, there may also be many differences between the genes of one mouse and another, and between populations of mice found at different locations. These differences within and between populations may be vital to the long-term viability of a species (**Freeland, 2003**). In terms of genetic diversity and conservation, there may be a domino effect during the decline of a species. As more individuals die, an increasing number of particular types of genes will go extinct. This leads to decreasing levels of genetic diversity within a species which, in turn, may reduce the longer term viability of the species (through an inability to adapt, say, to environmental changes) and so lead to more individuals dying and so on (see also **Hinchliffe and Belshaw, 2003**). Therefore, assessing within-species genetic diversity has become an important part of conservation biology. Even the elimination of populations of a relatively common species will lead to an overall reduction of biodiversity, because there will be an associated loss of many different types of genes.

## 2.2 How do we measure biodiversity?

Because it is such a multi-faceted concept, there is no such thing as a single measure of biodiversity. Attempts to quantify biodiversity will depend partly on the particular questions you are trying to answer, and partly on practical constraints. For example, the total genetic diversity of a population is called its 'gene pool'. But in order to completely measure the amount of diversity within the gene pool, we would have to sequence the entire **genome** of every individual within the population. Obviously this is not realistic – biologists can generally sample only a proportion of the population, and for each individual can investigate only a small region of the genome. Estimates of genetic diversity based on this approach will depend both on the individuals that were sampled, and the region of DNA that was characterized (some regions of DNA are much more variable than others).

The term '**genome**' refers to an organism's full set of DNA.

For a definition of DNA see **Freeland (2003)**.

As with genetic diversity, there is no straightforward way to measure species diversity. Species richness is defined as the number of species within a given area, but most measures of species diversity take into account the relative abundance of each species, i.e. how many individuals there are of each species. Imagine two samples of 100 individuals, each containing two species. Sample one contains 99 individuals of species a, and one individual of species b, whereas sample two contains 50 individuals of each species. Although both samples contain the same number of species, sample two would generally be considered to be more diverse.

Estimates of species diversity are also influenced by the sampling methods that are used, which will inevitably cause an unknown number of species (particularly the rarer species) to be overlooked (see **Morris, 2003b**, on sample measures). Furthermore, the numbers and types of species present will vary not only from year to year, but also within a year, for example between winter and summer. Surveying may be fairly reliable in areas of low diversity, or when surveying conspicuous creatures such as birds or large mammals, but it can be particularly limiting in highly diverse areas such as tropical rainforests, or when describing inconspicuous species such as soil invertebrates.

Perhaps the most difficult type of biodiversity to quantify is habitat diversity (see Figure 6.2). Imagine a stretch of river that includes waterfalls, riffles and pools. The depth and velocity of water will be very different in each stretch, as will

**Figure 6.2** The same stretch of river can contain many different habitats. These images show two different stretches of the River Tees in the UK.

numerous other variables including oxygen content, light, substrate and so on. As a result, different species will be found in different stretches of the river. Would you categorize this as a general river habitat, or would you further subdivide it into habitats of falls, riffles and pools? Within a pool, would you consider the edge to be the same habitat as the middle, even though it may be shallower and therefore home to a relatively greater number of plants? Or would you conclude that all rivers within a given area (e.g. in southern England) represented basically the same habitat? These are difficult decisions to make and yet they may be very important in questions of management because a greater variety of habitats leads to a greater diversity of species.

## 2.3  Does a growing awareness make a difference?

See Chapter Five for a discussion of global response to environmental issues.

In the 1980s, the *National Forum on BioDiversity* was just one of the events that contributed to a growing awareness about the threats to global diversity. In 1992 this awareness culminated in the *United Nations Conference on Environment and Development* (*UNCED*), also known as the Earth Summit, which met in Rio de Janeiro, Brazil, to discuss the global conflict between economic development and environmental protection. A major outcome of this meeting was the signing by more than 150 nations of the *Convention on Biological Diversity*. Article 6A of the convention required the signatories to develop national conservation plans for the sustainable use of biodiversity. In the UK, a major government response was to publish the UK Biodiversity Action Plan (BAP). The overall goal of the BAP was to prevent the further net loss of biodiversity within the UK. It currently outlines threats and opportunities for preserving 391 species and 45 specific types of habitat such as native pine woodlands and upland oakwoods.

### Activity 6.1

List the scientific and political developments of the last 20 years that make biodiversity conservation more likely today than in the recent past. What constraints have you heard about so far?

### Comment

The advent of conservation genetics along with the political and scientific establishment's attempts to place biodiversity on national and international agenda are opportunities for conservation and sustainable environmental development. Likewise, the Convention on Biological Diversity and national BAPs have helped to raise the profile of conservation. We have not heard about many constraints so far, but the uncertainties that surround the measurement of biodiversity may constrain conservation policies.

So, has global awareness of biodiversity and its threats made a difference? Does the future now look brighter for the many different life forms with which we co-exist? On the one hand, there are many examples of ways in which sustainable development has met various goals of conservation. For example, in 1994 the Ugandan government adopted a programme whereby revenue generated from wildlife tourism could be shared with local people, so giving incentives for biodiversity protection. Costa Rica's 1996 Forestry Law includes provisions to compensate private landowners and forest managers who maintain or increase the area of forest within their properties. And in different parts of the world, farmers are raising crops within mixed ecosystems. In Mexico, they are growing 'shade coffee' by putting coffee trees in a mixed tropical forest rather than in monocultural plantations that reduce biodiversity (**Morris, 2003a**). These farmers are then able to rely on natural predators found within the intact ecosystem, as opposed to relying on chemical pesticides. There are many more examples of local initiatives to merge sustainable economic development with the protection of biodiversity.

On the other hand, if we look at the global picture, more and more species are becoming extinct every year. The main reasons for this are often related to a combination of habitat loss, the spread of invasive species, pollution, inappropriate property rights, poor management, social inequality and, of more threat in the future, accelerated climate change (see **Morris et al., 2003** for more detail on many of these processes and issues). As a result, forests continue to be cut down, urban areas are spreading, agriculture covers increasingly large areas of land, coral reefs are dying and marshes are being drained (or drowned if they are on a coast susceptible to sea level rise). Human activity is at the core of these changes, and the human population is still growing (although the extent to which this growth will continue is a matter of debate and uncertainty). Apart from the growing numbers of humans, the intensity with which we are using up resources, developing unsustainable uses of habitats and polluting the atmosphere, seas and land is critical (**Reddish, 2003**). In the next sections we will look at responses to these unsustainable practices (Section 3) and review the likelihood of their success (Section 4).

## Summary

Biodiversity encompasses the entire range of variation in living organisms, from genetic variation to species and ecosystem variation. Although humans have been aware of this, either implicitly or explicitly, for thousands of years, it is only recently that we have seen international research and declarations aimed specifically at protecting biodiversity. It would seem that the concept of biodiversity has in many circles acted as a channel, allowing people to work towards a more unified goal. We could therefore argue that the coining of the term biodiversity, and the techniques of measurement that accompanied it, were important environmental responses to the threats facing habitats and species globally. In this sense, a change in thinking and knowledge is a major opportunity for sustainable development (aim one in our list on page 239). But we have also started to identify some major constraints (also aim one). There are some major challenges posed by the size and character of human societies that can make such optimism seem ill-founded. Finally, we have identified some of the sources of uncertainty that are encountered when trying to estimate biodiversity (aim three).

# 3    Valuing biodiversity

Every year, an enormous amount of money is spent on conserving biodiversity. Why are we so concerned with the continuing loss of life forms on our planet? The following quotation is from E.O. Wilson. As you read it you should note the kinds of values that Wilson is placing upon biodiversity.

> Why should we care? What difference does it make if some species are extinguished, if even half of all the species on earth disappear? Let me count the ways. New sources of scientific information will be lost. Vast potential biological wealth will be destroyed. Still undeveloped medicines, crops, pharmaceuticals, timber, fibers, pulp, soil-restoring vegetation, petroleum substitutes, and other products and amenities will never come to light. It is fashionable in some quarters to wave aside the small and obscure, the bugs and weeds, forgetting that an obscure moth from Latin America saved Australia's pastureland from overgrowth by cactus, that the rosy periwinkle provided the cure for Hodgkin's disease and childhood lymphocytic leukemia, that the bark of the Pacific yew offers hope for victims of ovarian and breast cancer, that a chemical from the saliva of leeches dissolves blood clots during surgery, and so on down a roster already grown long and illustrious despite the limited research addressed to it.
>
> In amnesiac revery it is also easy to overlook the services that ecosystems provide humanity. They enrich the soil and create the very air we breathe. Without these amenities, the remaining tenure of the human race would be nasty and brief. The life-sustaining matrix is built of green plants with legions of micro-organisms and mostly small, obscure animals – in other words, weeds and bugs. Such organisms support the world with efficiency because they are so diverse, allowing them to divide labor and swarm over every square meter of the earth's surface. They run the world precisely as we would wish it to be run, because humanity evolved within living communities and our bodily functions are finely adjusted to the idiosyncratic environment already created.

(E.O. Wilson, 1992, pp.330–1)

What is Wilson saying about the value of biodiversity? He is effectively saying that to lose biodiversity is to lose something of current value, and something that may well be valuable in the future. If we use the distinction discussed by **Hinchliffe and Belshaw (2003)**, Wilson is largely arguing that biodiversity has instrumental value for people in the form of raw materials, medicines and life-supporting processes. Yet, if biodiversity is so valuable, why is it under threat? We argue in this section that one reason for the threat is that this value is seldom realized by the people who determine the ways in which land, water and air are used. And a major reason for this is that the value of biodiversity is seldom expressed in the one language that seems to matter most to decision makers – money. Our argument is that one major opportunity for the sustainable use of biodiversity might be to find ways of valuing it in monetary terms.

# 3.1  Why revalue biodiversity?

Look at the following table, which lists the various **environmental services** or functions that forests provide. (An environmental or ecosystem service can be defined as the materials, processes and energy flows generated by an environment or ecosystem that humans find directly or indirectly useful. It is, you might note, an anthropocentric way of looking at environments).

environmental services

Table 6.1  Environmental services of forests

| Sources of materials and services | Sinks for wastes | General and life support |
| --- | --- | --- |
| timber | absorption of waste | genetic pool |
| fuelwood | recycling nutrients | climate regulation |
| other business products | watershed protection | carbon fixing |
| non-wood products | protecting soil quality and erosion resistance | habitat for people, flora and fauna |
| agricultural production | | aesthetic, cultural and spiritual source |
| recreation and tourism | | scientific data |

Source: Pearce and Moran, 1994, p.21.

## Activity 6.2

Using the services listed in Table 6.1, make two columns, one for a forest low in diversity (say a monocultural plantation of coniferous trees) and one for a highly diverse forest (say a deciduous lowland woodland in Europe, or a rainforest. For examples of both types of forest, see Figure 6.3). Under each heading list those functions that are best served by that particular forest, relative to the other forest. If you can't decide, leave the function to one side.

| Monocultural forest | Diverse forest |
| --- | --- |
| fuelwood | genetic pool |
| ... | ... |

## Comment

If you ended up with a table like the one at the end of the chapter you will have considerably more services in the diverse column than the monocultural column. This is an indication of the ways in which biodiversity can add value to an area. (Note the answers here depend on the details of the context, so you might have a table that differs from the one at the back of the book. The important issue is to have realized that forests with high biodiversity can provide some services more effectively than those with low diversity).

But which forest is worth the most in terms of money? To answer this, ask yourself which of the functions in your table you could most easily sell and therefore convert into money. As a generalization we would say that the monocultural forest, despite its limited number of ecosystem services, is actually of potentially higher value in monetary terms than the diverse forest. This is because many of the functions of the diverse forest are *external* to conventional economic analysis (they are positive externalities – externalities were discussed in Chapter 3). If you were a decision maker who wanted to maximize the value of your land, then, in conventional terms and assuming it was physically possible, it might pay to remove the diverse forest and cover the land with a monocultural plantation. It is worth noting, too, that not only would this make sense in terms of the functions for which monetary value can be realized but also there are often government subsidies for developing land in order to realize more economic potential. This tendency towards monocultural land use is, then, the result of two types of economic failure. The first is **market failure**, which is a failure to correctly value goods and services, or to inappropriately place value on some attributes more than others. The second is called a **government failure**, as it is governments that are often said to over-subsidize forestry, agriculture, and transport infrastructure by providing grants, tax breaks or preferential loans to these kinds of development.

market failure

government failure

**Figure 6.3** (a) A mixed deciduous woodland, high in biodiversity, and (b) a mono-cultural plantation.

Without actually saying so, we have mimicked a conservation/development contest – where it seems that there is more economic incentive to develop a monoculture than to conserve biodiversity. In order to confront this kind of issue a number of environmental and ecological economists have written about and advised governments and non-governmental organizations (NGOs) on the utility and means of calculating monetary values for otherwise undervalued or zero-priced environmental goods and services. In a study carried out for the World Conservation Union, Pearce and Moran argued strongly for expressing biodiversity values in terms of money in order that 'more powerful, more practical arguments can be formulated for its [biodiversity's] conservation' (Pearce and Moran, 1994, p.1). Many of the world's biodiversity hotspots (areas earmarked for their particularly high levels of biodiversity that are under threat from development) exist in parts of the world that are keen to develop economic returns for their natural resources. Valuing biodiversity in monetary terms may help in guiding those resource decisions towards more sustainable alternatives, allowing, at one and the same time, economic development and environmental conservation. In short then, re-valuing biodiversity is desirable in that it may facilitate sustainable use of environments and may help to generate sustainable income in poorer areas of the world. In the following sub-section we can look at the attempts of Pearce and Moran and others to value ecosystem services and biodiversity and consider some of the difficulties associated with their attempts.

## 3.2  The difficulties of valuing biodiversity in terms of money

Valuing environments in terms of money is a difficult process (see **Burgess, 2003**, for a discussion). Indeed, while it may be relatively straightforward to price some biological resources (like timber), defining a price for something as complicated as biodiversity is an uncertain process for at least four reasons:

1   the problem of indirect values
2   the absence of markets
3   problems of extrapolation
4   the problem of discounting

We will deal with each of these reasons in turn. First, unlike timber, which is directly usable, much of the value of biodiversity is indirect (see Box 6.1), which means that the benefits gained from the service are not for the service itself, but for some other service to which it contributes. This complexity makes it scientifically difficult to state precisely the role that the ecosystem is having (and would or would not have if it was altered), and economically difficult to find a reasonable estimate of the importance of that particular role relative to many other processes. Examples of indirect benefits of biodiversity would be watershed protection, climate regulation or recycling nutrients.

direct use value

indirect use value

## Box 6.1 Direct and indirect use values

**Direct use values** refer to actual uses such as fishing, timber extraction etc. They are often easily estimated by referring to current market prices for timber, fish etc.

**Indirect use values** refer to the benefits deriving from ecosystem services such as a forest's function in protecting a watershed. For further discussion of direct and indirect use values see Pearce and Moran, 1994. p.19 and **Burgess, 2003.**

While we can easily find a value for a forest in terms of its total timber – there are established costs of extraction and world prices for certain kinds of timber - biodiversity is difficult to price because markets do not exist for many of its services. For some economists unless you can actually realize a price on the market, then it is nonsensical to talk about a monetary value or price. This is our second area of uncertainty, listed on the previous page.

Nevertheless, another group of economists has attempted to estimate the value of the annual contribution of the world's ecosystems to the world's economy (note this is an attempt to value more than just biodiversity). The aim is to focus attention on the under-priced nature of eco-system services. Costanza and colleagues took over one hundred studies that had attempted to value ecosystem services in a number of biomes, and derived average annual values per hectare for each biome (Costanza et al., 1997).

see **Freeland (2003)** and **Morris and Turner (2003)** for a discussion of biomes.

These values were then used to calculate a worldwide value for ecosystem services by multiplying each per hectare value by the global area for each biome (see Figure 6.4). By adding all of these values together, the researchers estimated that the value of global ecosystem services was US$33 trillion per year (at 1997 prices). This figure is of similar size to global gross national product.

It should be noted that this study raises all manner of questions and is plagued by uncertainties. These reside not only in the initial studies (which relied on methods, like contingent valuation, that are themselves open to criticism, see **Burgess, 2003**) but also on the process of extrapolating micro-economic analyses up to the macro-economic level (our third area of uncertainty identified above). Such extrapolation can miss all the complexities of accounting for changes in value as goods and services become more or less available at a different spatial scale (Balmford et al., 2002). For example, the total global value of tropical forests may not be best calculated from a local estimate of value. In the local case study, from which the per hectare valuation is derived, there may be a relatively high percentage of forest as a proportion of local land cover, whereas when we look at the global situation this type of forest may make a very small percentage of the global land area. The result is that a valuation based on a locally abundant land use type may well be different to a valuation based on a globally scarce habitat. Simple extrapolation of local studies misses this change in value, and might result in a major underestimation of the total value.

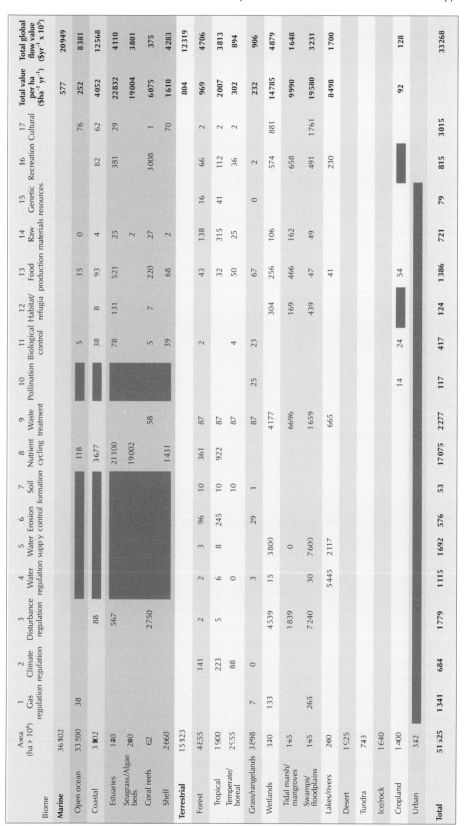

**Figure 6.4** A summary of the average global value of annual ecosystem services.

The numbers in the main body of the table are in US Dollars per hectare per year ($ha$^{-1}$ yr$^{-1}$). Row and column totals are in billion US Dollars per year ($yr$^{-1}$ x 10$^9$). Column totals are calculated by multiplying the per ha services in the table and the global area of each biome, and then adding all of these together. They are not the sum of the per ha services themselves. Shaded cells indicate services that do not occur or are known to be negligible. Open cells indicate lack of available information, which may mean that the total is a conservative estimate.

*Source:* Costanza et al., 1997, p.256.

Similarly, the value of a lake for recreation might be locally high, but when extrapolated might result in an over-estimation of total value given that there is likely to be a limit to the demand for such recreation globally. To suggest that Lake Baikal in Siberia has the same per hectare recreational value as Lake Windermere in Cumbria is an assumption that is worth questioning. These uncertainties, combined with the questionable idea of valuing the entire planet's natural capital and ecosystem services, may make this study sound foolhardy. Even so, it is fair to say that the figure of $33 trillion raised a good deal of media interest in the value of the earth's environment – drawing attention to services that we have hitherto taken for granted.

More recently, a related group of economists has taken a slightly different tack. In order to facilitate decision making that can compare the value of different uses of ecosystems and habitats these economists have focused not on estimating gross total values of biomes but instead have sought to estimate the difference in benefits that flow from biomes that are disturbed by human actions in different ways. So for example, the aim is not to compare the value of a forest *itself*, but rather to compare the value that the forest *provides* in terms of goods and services over time with the value that the same area would provide over time if it was converted to, for example, farmland, car parking or housing. Box 6.2 gives a real example of such an analysis, but before you read this, we need to review our fourth area of potential uncertainty concerning economic valuations of biodiversity: the problem of discounting.

The main uncertainty involved in comparing income streams over time is that a lot depends on how we take account of change in the value to us of money over time. As you saw in Chapter One, discount rates are often used in order to adjust current values to those that might apply in the future. In the radioactivity example the issue was the distribution of the costs of dealing with radioactive waste through time (who pays and when?). In our case, we are more concerned with the distribution of the benefits of using an area of land through time. Some land uses will deliver almost immediate benefits whereas for others we may have to wait for more than one generation to see the rewards of our actions. Conventionally, it is believed that people would rather have benefits now than wait 20 years for them to accrue. If someone offered you £1,000 today, or the option of having £1,000 in twenty years time, what would you do? Would you go for the money now? Some would take the money because they believe they will be wealthier in the future (so making £1,000 less significant in terms of their overall income in 20 years time). Others will want to enjoy spending the money sooner rather than later as they don't know what the future will hold. All this conspires against the future, so that, in an economic analysis, it often makes sense to discount the future (or calculate this loss of benefit through time). Typical discount rates lie in the range of 4–10 per cent per annum (effectively meaning that the monetary value of something is reduced by between 4 and 10 per cent of its value at the beginning of the year). The higher the discount rate, the more quickly value is assumed to decrease and, therefore, the more the present is being favoured over the future (or the more likely that you will take the money now rather than wait for any future benefits). People disagree on the extent to which we should value or discount the future. Some environmentalists suggest that in order to properly value the future, and so make sustainable decisions, we need to adopt low or even negative discount rates. Others are more optimistic and suggest that we need to sort out so many issues in the present that the future can be effectively discounted (see also Chapter One for different principles regarding the rights of the future). The outcome of these debates has practical effects. Faced with two forms of development, one that offered small benefits in a short period of time and another that offered large benefits but many years from now, our decision on which to take would most likely depend on the discount rate being used.

## Box 6.2  Revaluing biodiversity to enhance sustainable decision making

The following extract has been rewritten from Balmford et al. (2002). The Balmford article was based on work originally reported in Yaron (2001) and compares the benefits of three different types of land use in Cameroon over the next 32 years: reduced impact logging (a form of sustainable forestry); small-scale agriculture (involving clearing part of the forest but retaining large areas); and plantation (involving removing the forest and replacing with rubber and oil palm monocultures). The study involves a discount rate assumed to be 10 per cent. Read the extract, which relates to Figure 6.5, and then answer Activity 6.3.

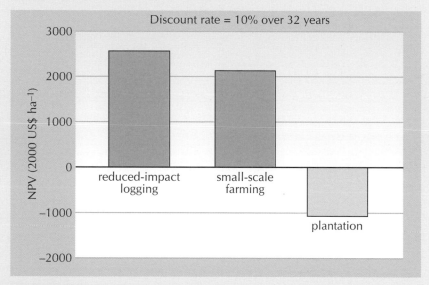

Figure 6.5  The benefits of retaining and converting natural habitats, expressed in NPV (net present value). The three land-use options are: reduced impact logging; small-scale agriculture; and plantation. The calculations are based on the benefits of the three schemes over the next 32 years.

**Figure 6.6** A closer view of the forest habitat near to Mount Cameroon.

A study from Mount Cameroon, Cameroon (see Figures 6.6 and 6.7a and b), compared low-impact, sustainable logging with more extreme land use changes (including conversion to plantations of oil palm and rubber trees, and conversion of forest to smaller-scale farming enterprises). Findings suggested that private interests would benefit most from conversion of the forest to small-scale agriculture. A second alternative to retaining the forest, the conversion to oil palm and rubber plantations, which looked attractive to investors, in fact yielded negative private benefits (or more costs than benefits) once the effect of market distortions had been removed.

Meanwhile, social benefits (those benefits enjoyed more widely than private interests) from non-timber forest products (NTFPs), sedimentation control, and flood prevention, were highest under low-impact, sustainable logging. In addition, this sustainable form of forestry yielded

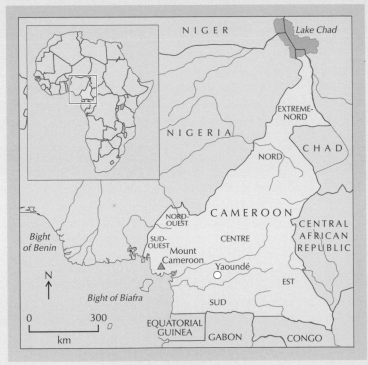

**Figure 6.7(a)**    a location map of Cameroon, showing Mount Cameroon.

**Figure 6.7(b)**   An aerial view of the Mount Cameroon rainforest.

most benefits globally in terms of carbon storage and in terms of values held by people living outside the region. Overall the Total Economic Value (TEV) of the sustainable form of forestry was, taking all these private, social and global benefits into account, 18 per cent greater than that of small-scale farming (around $2750 compared with $2110 ha$^{-1}$), whereas plantations yielded a negative TEV.

*See* **Burgess (2003)** *for a discussion of total economic value (TEV).*

Extract based on Balmford et al., 2002, p.950–1

## Activity 6.3

In the light of the discussion in Box 6.2 and the discussion of discounting above, answer the following questions:

1   Of the three developments, and assuming conventional forms of valuation, which would be favoured by the private interests?

2   What would happen to the land if the total economic value was taken to guide decision making?

3   If most of the benefits of sustainable forestry accrued after year 20, and most of the benefits of the plantations accrued in the first 20 years, what effect might an increase in the discount rate have on the decision? What about a decrease in the discount rate?

### Comment

1   Private interests would make most money from small-scale farming, but plantations might result even without positive private benefits if market distortions were sufficient.

2   If valuations were based on TEV then we would expect sustainable forestry to develop as it would yield the highest income.

3   Increasing the discount rate would value the present more than the future, with the result that plantations, with their more immediate returns, would become more likely. A decrease in the discount rate would make the sustainable forestry option less unattractive to an investor.

Re-valuing biodiversity, or other components of environments, is essentially an attempt to correct the failures of current market and government policies, which tend towards short-term gains with little consideration given to environmental quality (see also Chapter Three). This may provide a vital opportunity for guiding markets and governments towards sustainable development. In the past, pollution, resource depletion and degradation were regarded as, perhaps, the inevitable consequence of economic growth. However, in recent decades, a new discourse of development called **ecological modernization** has emerged, a discourse that suggests it is both possible and desirable to develop economically and socially while, at the same time, conserving environments (discourse is defined in **Hinchliffe and Belshaw, 2003**). The forms of valuation that we have discussed in this section so far are part of this shift towards finding ways of meeting current economic requirements, environmental quality objectives and the duty to respect future generations' needs. We can now elaborate upon this development discourse.

ecological modernization

## Summary

This section has focused on identifying some of the opportunities that exist for sustainable development. In particular we have focused upon the opportunities that exist for re-valuing biodiversity (and other aspects of environments), and in so doing, steering resource decisions away from non-sustainable uses (aim one in our list on page 239). We have also discussed some of the economic uncertainties that exist when trying to revalue environments – they are bound up with indirect values, non-market values, problems of extrapolation and discounting the future (aim three).

# 4   The discourse of ecological modernization

Ecological modernization has become a dominant discourse of sustainable development. It supports an optimistic interpretation of contemporary social development. Initially, ecological modernization was a term used to describe the ways in which western industrialized societies, especially those in north-western Europe, incorporated environmental factors into decision making. There are four aspects to this:

*   The first is *technological*: the application of technical controls to production processes. This has been called the 'refinement of production' (Mol, 1995) –

protecting the environment using pollution controls, and conserving resources through waste minimization, reuse and recycling. A whole battery of technologies and techniques have been applied to a wide variety of industrial processes and examples of some of these, for instance life cycle analysis, were discussed in Chapter Two. The application of technological controls not only benefits the environment but often contributes to efficiency as companies able to meet pollution standards are also able to consolidate their competitive position (see Chapter Three).

- The second aspect is *political*: linking ecological modernization to the fundamental changes that occurred in politics and markets in western democracies in the last quarter of the twentieth century, and which is sometimes referred to as 'political modernization'. This recognizes and reflects the primacy accorded to the market as the means of resource allocation and decision making. The market is widely regarded as flexible, efficient, responsive and innovative. The introduction of economic instruments as means of reducing pollution (discussed in Chapters Three, Four and Five) is a form of response that fits neatly into the discourse of ecological modernization, as are attempts to place monetary values on previously under- or zero-priced environmental goods and services (see Section 3 here and **Burgess, 2003**).

- The central role of markets in decision making has led to a third aspect: the reshaping of the relationship between the *state* and *society*. There has been a marked shift in emphasis from state control to the notion of an 'enabling' state engaged in co-operative relationships with business. The emphasis is on negotiation and self-regulation as a way of ensuring the development of mutually acceptable goals and regulations. The state's role is, to some extent, more residual, providing a framework of legislation, targets and indicators with supporting monitoring, inspection and (as a deterrent) enforcement regimes.

- A fourth aspect is the part played by *civil society* under conditions of ecological modernization. **Civil society** includes organized groups and associations that operate independently from government and the state. Among these are environmental groups including NGOs and less formal, citizens-based environmental movements. NGOs and citizens are encouraged to participate in various forms of consultation, dialogue and so-called 'deliberative' approaches to policy making to achieve widespread social consensus (see Chapter One). Under ecological modernization they are seen in a collaborative role with business and government thereby supporting and legitimizing environmental policy making.

*civil society is also defined in Chapter Four*

## 4.1 Ecological modernization in a country in transition

Ecological modernization, from its origins as a description of conditions developing in western Europe, has become a broader discourse reflecting approaches to sustainable development now being engaged by capitalist societies in a wider variety of contexts (see **Castree, 2003**, for a definition of capitalism). One such context is in the former communist countries which, since the revolutions at the end of the twentieth century, have come to be called 'countries in transition', as they move from being centrally planned to market economies. They illustrate how the ideas and practice of ecological modernization have been embraced during a period of great economic and political uncertainty. In such conditions the strengths and weaknesses of ecological modernization as a way of managing environment and development are clearly revealed.

The case of Lake Baikal in eastern Siberia (see Figure 6.8a) provides a good example of the process of ecological modernization in a country in transition. The lake is over a mile (1.6 km) deep, 400 miles (640 km) long and up to 50 miles (76 km) wide covering an area the equivalent size of a country such as Austria and the length of England (see Figure 6.8b). It contains a fifth of the world's unfrozen fresh water. The lake provides an environment incredibly rich in biodiversity and, of the 2,500 or so species found in the lake, about three-quarters are endemic. The international importance of the lake's environment and its surroundings has been recognized with its designation as a world heritage site by UNESCO in 1996 (Figures 6.9 and 6.10).

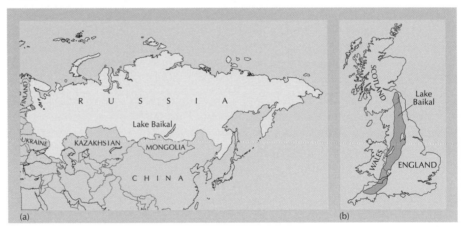

**Figure 6.8** Lake Baikal (a) the location of Lake Baikal in Russia; (b) superimposed on Britain, Lake Baikal would stretch from Cornwall to the Scottish borders.

Yet this is an environment potentially under threat from degradation and pollution. The construction of a hydroelectric power dam on the Angara, the only outflowing river, in 1959 raised the level of the lake, causing flooding. Logging in the coniferous forests (taiga, see Figure 6.11 overleaf) surrounding the lake has caused soil erosion and there is pollution from agricultural, tourist and urban development especially from the city of Ulan-Ude and down the Selenga (the main river that enters the lake). At the northern end of the lake the construction of the Baikal-Amur Mainline (BAM) railway and associated construction settlements caused soil erosion, siltation and pollution in the lake and its surrounds.

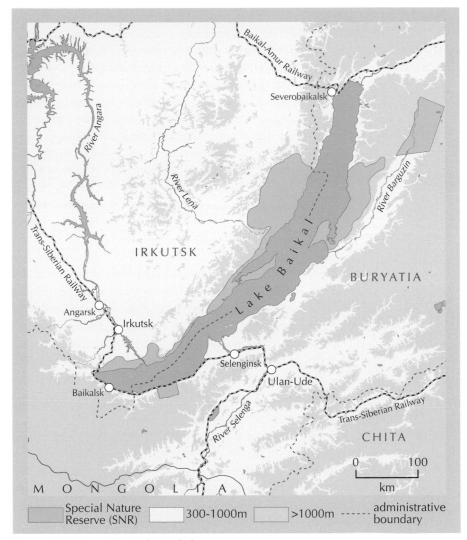

**Figure 6.9** Map of the Lake Baikal area.

**Figure 6.10** Lake Baikal (a) is designated as a world heritage site by UNESCO and contains around two and a half thousand species, including (b) the Nerpa seals, the only freshwater seal in existence.

**Figure 6.11** The Siberian Taiga stretches for hundreds of kilometres and consists of birch, larch and pine trees.

But the most visible threat is industrial pollution from coal burning plants, aluminium smelting and above all from the pulp and paper mills (PPM) at Selenginsk and Baikalsk (Figures 6.9 and 6.12). The Baikalsk PPM was constructed in the 1960s, initially for military purposes producing cellulose for high quality aircraft tyres. From the outset it was controversial, attracting opposition from scientists and environmentalists who were concerned about pollution and health threats from dioxins, sulphates, chlorides and persistent organic pollutants discharged into the lake and the atmosphere. While there is continuing scientific debate as to just how palpable the threat to the lake may be, environmentalists have argued that the risk to the lake's ecosystem is too great and therefore the plant should be shut down. But shutdown poses a severe economic risk to Baikalsk, a town of around 30,000 people and an example of a 'peripheral community' (referred to in Chapter Four and defined in **Blowers and Elliott, 2003**), isolated and largely dependent on the mill and with few alternative sources of employment.

The Baikalsk PPM demonstrates the basic ingredients of ecological modernization. During the Soviet era and subsequently, the plant was technologically modernized in response to the requirements of environmental legislation. The production process was upgraded to reduce pollution, treat water and conserve energy. At the time of *perestroika* (restructuring) in 1987, with environmental concerns gaining ground, plans to re-profile the plant were accepted though not implemented. But, by the turn of the century with a federal law on the protection of Lake Baikal in place, a World Bank grant had been secured for the introduction of a closed-loop water system to eliminate discharges into the lake. With the plant now privatized and adopting market criteria, the costs of environmental improvements (estimated at 8 to 14 per cent of total costs) had to be covered by improved productivity and efficiency if the plant was to remain competitive. In terms of its relations with the state, the plant has accommodated to the need to secure its markets and profits while, at the same time, achieving acceptable environmental standards through negotiation with state authorities. Given its location in a particularly sensitive environment the plant has been under close scrutiny from environmental groups that have played a key role in ensuring the protection of the lake. In short, the Baikalsk PPM has modernized to meet environmental objectives while sustaining industrial production.

**Figure 6.12**   View of Baikalsk pulp and paper mills.

The case of the Baikalsk PPM suggests that through ecological modernization there are significant opportunities for sustainable development. Ecological modernization assumes that through co-operative relationships between business and the state, environmental protection and economic goals of profit and growth are compatible. Yet there is room for doubting this comfortable thesis as we shall see in a moment.

## Summary

In this section we have described a discourse called ecological modernization, which considers environmental protection in relationship to economic criteria. We have seen, through the Lake Baikal case study, that opportunities exist for forms of modernization or progressive social change that do not necessarily undermine environmental quality. In response to the first aim of this chapter, on page 239, we have outlined a number of *opportunities* for sustainable development. And with regard to the fourth aim, we have started to construct an optimistic view of the future, in which, it is argued that environmental and economic goals can be achieved simultaneously. In the next section we consider the *constraints* on sustainable development, by looking at some of the political uncertainties that relate to the implementation of some of these modernizing attempts.

# 5    Biodiversity, ecological modernization and 'business as usual'

So far we have introduced biodiversity and discussed efforts to place monetary values upon it. We have also introduced the notion of ecological modernization to describe a prevalent assumption that it is possible to square the circle of development and environmental conservation. If you were an optimist, pure and simple, you might stop there and suggest that we now understand what biodiversity is, its measurement, how to value it properly and therefore how to use it effectively. More than that, you might conclude we have an overarching sense, through ecological modernization, that we can build wealth at the same time as conserve or enhance natural capital and environmental services.

In this section we examine the grounds for taking a more pessimistic perspective. First, we consider ecological modernization in a wider context, one that leaves us with fewer grounds for optimism. Then we examine some of the concerns that people express over the ways in which biodiversity is being taken up and used. We do so by looking at a paradox of biodiversity conservation – relating it to the role of food industries in promoting biodiversity use and conservation.

## 5.1  Ecological modernization – neglecting environment and development

### Activity 6.4

Can you think of reasons why, in a broader context, ecological modernization may not achieve sustainable forms of development? Consider this question in terms of the four aspects of ecological modernization identified in the previous section. When you have thought about it return to the discussion below.

Ecological modernization originated as a limited description of political and technological responses to ecological concerns in advanced countries in western Europe. It has been translated into a discourse of sustainable development relating to a much wider range of technologies and political systems. Yet its initial limitations still apply and undermine its claims to universal application. We can explain this by returning to the example of Lake Baikal, this time looking at the lake in its broader regional and international context. We shall take each aspect of ecological modernization in turn.

*Technological.* In terms of technology ecological modernization can be readily applied to those industries, like pulp and paper making, where processes can be modified to reduce resource use, waste or pollution. But, there are many activities that cannot be modernized in the sense of removing their threat to the environment. They include, for instance, technologies whose impacts are uncertain (such as GM crops – see **Bingham, 2003** and **Bingham and**

**Blackmore, 2003**), those where technical solutions are unproven (e.g. radioactive waste) and those whose environmental impacts are difficult, if not impossible to reverse (such as extensive logging and exploitation of other resources that create degradation and pollution – see **Morris, 2003a**). Seen in this broader context, ecological modernization has a relatively limited application. In the case of Lake Baikal the environmental impacts of logging, oil pipelines, minerals exploitation, railway construction and other large-scale activities are more likely to damage the ecosystem than a pulp and paper mill whose pollution can, ultimately, be controlled or prevented.

*Political.* Ecological modernization assumes that environmental and economic concerns are compatible. However, as we saw earlier and in Chapter 3, the market is not very good at placing appropriate monetary values on environmental goods and the tendency is to undervalue or even to ignore them. Even when we have a fairer price, the tendency is for markets to focus on short-term signals and thereby neglect the longer-term risks to the environment (as is evident in our discussion of discounting earlier). In the case of the Baikalsk PPM the need to achieve profitability in the new market conditions of post-Soviet Russia was paramount and environmental improvements became a burden that could only be met by government subsidy and international financial assistance. When economic and environmental objectives come into conflict, it is very likely that economic ones will prevail. The evidence from elsewhere in the former Soviet Union (and in many other countries across the world) suggests that long-term, possibly irreversible environmental damage can occur as a result of meeting short-run economic demands (see Figures 6.13 and 6.14 overleaf).

*State and Society.* Ecological modernization looks to the state to provide a framework of environmental policy and regulation which it develops in negotiation and partnership with business interests. This relationship ensures that regulation will respond to economic criteria. Implementation is in the hands of business through self-regulation and government agencies that are responsible to ensure compliance with environmental standards. There is sometimes an implementation deficit, which can be substantial in a country such as Russia where controls are weakly enforced and corruption is a considerable problem. During the transition to a capitalist economy the decentralisation and dispersal of regulation has weakened the state control of the former Soviet Union. In addition, the weakness of the Russian economy has reduced the pressure for environmental control. As industrial production has declined it has become easier to meet targets so that a situation of 'compliance without implementation' occurs (KPMG/CESAM, p.73). Looked at in this context the effectiveness of regulation at Lake Baikal is exceptional. Ecological modernization also emphasizes the political level of the nation state and so neglects the international dimension of environmental policy making. As the case of Lake Baikal indicates, transboundary pressures on the environment are intensifying. With the massive populations and markets in China, Japan and other Far Eastern countries on its borders, Siberia is wide open to resource exploitation. China satisfies nearly half its timber needs from Russia and there are plans for oil pipelines that would cross the Lake

**Figure 6.13** The former Soviet Union has its ecological disasters: (top) industrial pollution from steel works near Magnitogorsk, in the Urals and (bottom) diversion of water for irrigation has converted much of the Aral Sea into a desert.

Baikal region linking Siberian oil fields to massive new settlements in China. The resulting environmental degradation and possible pollution will have impacts on ecosystems that are possibly beyond the scope of ecological modernization.

*Civil Society.* Lake Baikal is also exceptional in that it was a focus of environmental concern and protest in Soviet times and has remained so ever since. Although protest has proved effective in helping to ensure strong protection for the lake and the modernization of the Baikalsk PPM, there is little evidence that environmental

**Figure 6.14(a)** Some of the Russian Taiga has been degraded by the effects of pollution.

movements are routinely consulted or have become incorporated into the policy-making process. For a while, during *perestroika*, they played a formative part in reforms leading ultimately to the dissolution of the Soviet Union. During the transition the partnership between business and the state has been maintained but there has not been much power extended to the environmental movement. Control remains firmly in the hands of a political and business elite.

Looked at in a wider context, ecological modernization is revealed as limited in application. It is limited to those activities where effective environmental controls are feasible; it assumes market criteria can be applied to environmental problems; it presumes a consensual basis for decision making whereas fundamental inequalities persist; it endorses political systems that are elitist and exclusive. By providing an optimistic analysis of environmental futures, ecological modernization is a beguiling discourse. But, if we take account of the technological, political and social concerns that are neglected by ecological modernization we may perceive an altogether more pessimistic perspective of our environmental future.

We can now apply this more pessimistic reading of ecological modernization to the biodiversity issue more generally.

**Figure 6.14 (b)** A view of the Chernobyl power station, which has caused widespread damage to ecosystems and to health.

## 5.2  Biodiversity and bio-uniformity – the paradoxical role of the bio-industries

In this subsection we are going to analyse the extent to which attempts both to economically develop and to conserve areas high in biodiversity are meeting with success. In other words, how successful are those efforts that aim to ecologically modernize human use of biodiversity? Our example will be the use of biodiversity as a resource in the production of new plant varieties and species for the food industries. First, we need to banish a common misunderstanding that relates to biodiversity conservation. Human beings are not, of necessity, bad for biodiversity. Indeed, in many cases and especially in terms of habitat diversity, human actions can enhance biodiversity. Certain livelihoods and forms of human habitation have, over many years, established unique ecologies, helped in the diversification of habitats and species and, through human selection, enhanced the expression and reproduction of certain genetic traits. While the global balance is almost certainly towards human actions reducing biodiversity, there are nevertheless countervailing tendencies whereby humans have had a hand in conserving and sometimes enhancing diversity (see Figure 6.15). One problem that is currently faced in biodiversity conservation is that many of the practices and livelihoods that have, in the past, led to some form of biodiversity conservation or enhancement, are themselves under threat of extinction. Some of these threats are, paradoxically, from the First World conservation establishment, which has often imagined that biodiversity is best served by keeping human beings at bay. As the environmental historian, William Cronon, puts it:

> ... the convergence of wilderness values with concerns about biological diversity and endangered species has helped produce a deep fascination for remote ecosystems, where it is easier to imagine that nature might somehow be 'left alone' to flourish by its own pristine devices. The classic example is the tropical rainforest, which since the 1970s has become the most powerful modern icon of unfallen, sacred land – a veritable Garden of Eden – for many Americans and Europeans. And yet protecting the rain forest in the eyes of First World environmentalists all too often means protecting it from the people who live there. Those who seek to preserve such 'wilderness' from the activities of native people run the risk of reproducing the same tragedy – being forceably removed from an ancient home - that befell American Indians. Third World countries face massive environmental problems and deep social conflicts, but these are not likely to be solved by a cultural myth that encourages us to 'preserve' peopleless landscapes that have not existed in such places for millennia. At its worst, as environmentalists are beginning to realize, exporting American notions of wilderness in this way can become an unthinking and self-defeating form of cultural imperialism.

> (Cronon, 1996, p.81–2)

Biodiversity cannot be conserved without making sure that people and livelihoods are also conserved. This is a bald statement, and one that we might take from Cronon's writing. It is also suggestive of the need, in producing a conservation strategy, to consider human ecology and economics as well as non-human life. In other words, given that we should be concerned with development and environment, this may be an area where the discourse of ecological modernization is applicable. But the risks faced by people are not just from well-meaning conservationists who would exclude people from their environments. They can also stem from the ecological modernizers' attempts to value biodiversity by turning it into a marketable resource. In order to find out how, we can look at the example of biodiversity and food production.

> Cronon's points linking particular, and archaic, views of environments to sometimes misguided conservation practices are echoed in **Belshaw (2003)** and **Morris (2003a)**

Some of the forms of biological diversity that have been fostered and nurtured by human activities are those that are most valuable to others. For example, varieties of plants cultivated in agricultural systems can be of great interest to producers of food and medicines elsewhere as they may contain particular properties that can be used directly or, through the biotechnology industries and genetic modification technologies, in the production of new varieties and hybrids (see **Bingham, 2003** for a discussion of genetic modification). This is in part a reason for the high potential instrumental value of biodiverse habitats – they contain the raw materials for food and other production technologies. As much of this resource is as yet unknown, there is a strong case for the conservation of biodiverse habitats and the livelihoods that have in part produced the biodiversity.

But there is a paradox – the development of marketable food products from biodiverse habitats or from diverse forms of agriculture may end up destroying the very thing that makes these forms of biodiversity possible in the first place. To find out more about this risk read the following three extracts, and then move on to the activity that follows.

**Figure 6.15** Two examples of intercropping (farming practices that use more than one crop): (a) here, in Cuba, bananas and corn are planted alongside each other, (b) this area, in Granada in Spain, is being used to grow a mixture of potatoes, olives and almonds.

Current international activities surrounding the genetic resources of plants aim to confront one paradoxical problem. This is that scientists throughout the world are rightly engaged in developing better and higher yielding cultivars of crop plants to be used on increasingly large scales. But this involves the replacement of the generally variable, lower yielding, locally adopted strains grown traditionally, by the products of modern agriculture – the case of uniformity replacing diversity. It is here that we find the paradox, for these self-same plant breeders are dependent upon the availability of a pool of diverse genetic material for success in their work. They are themselves dependent upon that which they are unwittingly destroying.

(Ford-Lloyd and Jackson, 1986, p.1)

The main problem with viewing biotechnology as a miracle solution to the biodiversity crisis is related to the fact that biotechnologies are, in essence, technologies for the breeding of uniformity in plants and animals... Although breeders draw genetic materials from many places as raw material input, the seed commodity that is sold back to farmers is characterized by uniformity... The exploitation of genetic diversity for crop improvement should be the ultimate objective of the exploration and conservation of genetic resources, it is argued. The arbitrary inequality created in the status of germ plasm creates an arbitrary separation between production and conservation. Some people's germ plasm becomes a finished commodity, a 'product', other people's germ plasm becomes mere 'raw' material for that product. The manufacture of the 'product' in corporate labs is counted as production. The reproduction of the 'raw' material by nature and Third World farmers is mere conservation. The 'value added' in one domain is built on the 'value robbed' in another domain. Biotechnological development thus translates into biodiversity erosion and poverty creation.

(Shiva, 1995, pp.206–7)

Germ plasm in this context refers to the material that contains the genetic diversity of a plant population. As a term it has generally been superseded in recent years.

The destruction and loss of biodiversity is closely linked to the loss of control by communities of their resources. Many factors have contributed to the loss of control farmers and communities have faced over the past few decades. The promotion of commercial seeds, such as so called 'high yielding varieties (HYVs)' and more recently genetically modified crops, has put the control firmly in the hands of the larger companies and institutions that develop the seeds. Using these seeds often demands the use of external inputs such as pesticides and fertilizers, again taking control out of farmers' hands. Credit for farmers is often conditional on the use of external inputs.

Agricultural research has also been biased towards the use of external technology – very little research involves the farmers and communities themselves. Such top-down research often ignores local cultures, traditions, diets and environments and results in seeds and practices that fail to live up to their promises. It also ignores and belittles the extensive knowledge that farmers have, and turns them into production workers rather than researchers and decision makers. Many families and communities feel a growing need, but decreasing capacity, to regain or retain this control over their farms and the genetic resources that they depend on.

(GRAIN, 2002, p.17)

## Activity 6.5

Using the extracts above, answer the following question: why might crop development, using HYVs and GM, be unsustainable?

## Comment

One reason is that farmers who have been instrumental in the production of biodiversity will be encouraged to adopt the new products of biotechnology. This produces a high risk of halting the very activities that nurtured the biodiversity and, instead, promotes a more uniform global agriculture. As these new practices are adopted, knowledge of the older ways of farming will be lost.

A second reason is the imbalance of power that accords value to some activities (those of the laboratory and western scientists) and little or no value to others (those of the farmers). Therefore, biodiversity only becomes valuable after it has been processed and its value realised on the global market of seed manufacture and trade. Little or no value is accorded to the local people (or to their farming practices) – the result being that the very activities that create biodiversity in the first place are under threat of extinction.

Taking these two together, farmers who see no value given to their work of nurturing biodiversity will be pressured to adopt the new varieties and species available from the biotechnology and seed companies and so contribute to biodiversity decline.

The inequalities in power that we have touched on here make for an uncertain and risky political climate when it comes to using the potential value of indigenous biodiversity as a means to sustainable development (and as a form of ecological modernization). There are many more aspects to this political scene – including the imbalances that are met when farmers adopt new seed varieties that have **terminator genes** engineered into them. Terminator genes have the potential to prevent plants producing viable seed, with the result that each year farmers will be forced to buy seeds from the manufacturers rather than cultivating their own. From the perspective of the manufacturer, this has the effect of protecting the seed variety, which may have eaten up years of research and development. It is a form of copyright protection. From a biodiversity perspective there are two possible arguments. On the positive side, it may reduce the risk of genetically modified species breeding with wild species (see **Bingham and Blackmore, 2003** and **Bingham, 2003**, for the possible consequences of this interbreeding). On the negative side, it brings to a halt the human practices of breeding and selecting that have, down the years, created a relatively rich variety

terminator gene

of crops and genetic resources. While there is currently an international moratorium on terminator technology, the effect of producing a relationship of dependency between farmers and seed companies is in fact widespread. Similar dependency results from patents (taken out by companies) that transfer property rights to the company and make it a legal requirement that farmers desist from producing their own seeds and/or distributing any surplus seed to other farmers. These technological and legal developments not only lock the farmers into a relationship with large companies, they also, as the last extract above mentioned, steadily erode the skills of those farmers in the selection, breeding and sowing of new varieties. In short, those lifestyles and people who have, often inadvertently, conserved and enhanced biodiversity are also at risk and are becoming a dying breed – their knowledge and practices are threatened as their dependency on large corporations grows. This erosion of biodiversity is a major casualty of a huge disparity in political and economic power between the North and the South. Figure 6.16 is a schematic representation of the production of biouniformity.

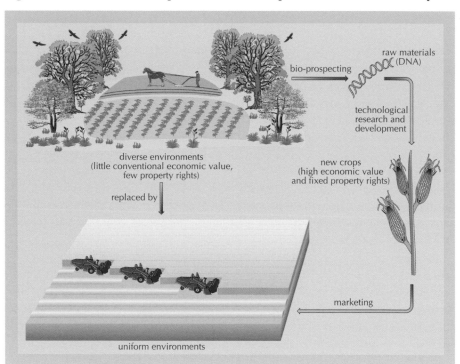

**Figure 6.16** The production of biouniformity and cultural uniformity

To conclude this subsection, we have now expressed a number of reasons to be less optimistic about the future of biodiversity. Despite our increased knowledge and the willingness and methodologies to revalue the economic worth of diverse habitats, there are some potential problems. We have argued that in many cases people are the custodians and reproducers of biodiversity but the very activities that have helped to generate biodiversity are under threat. Modernization here does not, even in ecological guise, address the massive inequalities in terms of access to economic, legal and political resources, or the ability to shape discourses, that exist most evidently between the global food corporations and the small-scale farmers. The paradox that may yet undo this imbalance is that the powerful global corporations may end up threatening the activities that made their biodiversity prospecting possible in the first place. Whether this will be enough to generate change in the ways in which biodiversity is economically valued and marketed may well depend upon who or what has the power to put forward alternatives. Currently state governments of the North seem loathe to check the global seed corporations and those of the South are often powerless to do so. Both sets of governments are seemingly easily bought by the promise of new revenue and the prestige that biotechnology offers. They often share with the corporations the desire for quick returns on investment – a short-term approach that can afford to ignore the paradox that we just raised, and that does not bode well for biodiversity conservation.

## Summary

In this section we have focused on the constraints that affect the future of sustainable development (the first aim in our list on page 239). Ecological modernization was seen to have limits in terms of its applicability. These included technical limits (application of technology is not always benign and there are industries to which it cannot be applied), political limits (markets are not particularly good at serving longer term aims), limits on the power of states (particularly when states are rendered less powerful by internal problems and international pressures), and limits to the effectiveness of civil society (especially where powerful business interests hold sway). These limits have been applied to the case of Russian modernization but have also been raised, in a different form, with respect to the use of biodiversity for global food production. We argued that the development of new species and high yielding varieties from the stock of global biodiversity may, through the global distribution of the products, undermine the very activities that created that diversity in the first place. This section has described how huge inequalities between and within nations, and between bio-industries and farmers, suggest that the future of biodiversity, which relies on the sustainability of a diversity of livelihoods, may be in the balance (the second aim in our list).

# 6    Is there hope for biodiversity?

Given the powerful forces described in the previous section, can biodiversity be conserved and even enhanced? Before we draw out some general themes that enable you to assess this issue in both optimistic and pessimistic terms, we will look to one more issue that relates to biodiversity. It is the possibility that the term 'biodiversity' and other terms that celebrate the richness of living and nonliving environments can be clawed back from the globalizing tendencies that you have read about and can be used to raise people's awareness of and capacity for actions in their surroundings. William Cronon captures this project nicely when he talks about his and Leopold's surroundings:

When I think of the times I myself have come closest to experiencing what I might call the sacred in nature, I often find myself remembering wild places much closer to home. I think, for instance, of a small pond near my house where water bubbles up from limestone springs to feed a series of pools that rarely freeze in winter and so play home to waterfowl that stay here for the protective warmth even on the coldest of winter days... And I think perhaps most especially of the blown-out, bankrupt farm in the sand country of central Wisconsin where Aldo Leopold and his family tried one of the first American experiments in ecological restoration, turning ravaged and infertile soil into carefully tended ground where the human and the nonhuman could exist side by side in relative harmony. What I celebrate about such places is not *just* their wildness, though that certainly is among their most important qualities; what I celebrate even more is that they remind us of the wildness in our backyards, of the nature that is all around us if only we have eyes to see it.

Indeed my principal objection to wilderness is that it may teach us to be dismissive or even contemptuous of such humble places and experiences.

(Cronon, 1996, p.86)

Cronon is no doubt right, in this and the earlier extract, to raise the concern that the way in which biodiversity is presented can reproduce the tendency of some environmentalists and people more generally to ignore the everyday and the local. This makes the optimism with which we greeted the invention of the term more suspect. One major shortfall of the UK steering group's BAP report, for example, is that its focus is upon the rare and the relatively distant (in terms of proximity to human populations) and that the consideration of urban environments is given short shrift. The political importance of urban human populations to the development of environmental consciousness and action is obvious. Most

**Figure 6.17** These leaflets describe small nature conservation schemes throughout the UK.

of the world now live in cities (see **Drake and Freeland, 2003)**, so ignoring urban wild spaces, and people's gardening and other practices, is a missed opportunity. The term biodiversity may also be too 'scientific' to capture the range of emotions and concerns that people have for these and other environments. But whatever name is given to wildlife, nature or environments, the variety of do-it-yourself, make and mend attempts to change local surroundings for the better can sometimes be a source of hope (see Figures 6.17 and 6.18). The following extract evokes one such scheme.

> Between the neighbouring Kings Cross and St Pancras Stations in the heart of London, a noisy and dusty street leads northwards between Victorian walls, warehouses and gasometers. On the left side of Camley Street a refuse collection site emits deafening noise and clouds of dust. Opposite, behind substantial wooden palings, between the street and the Grand Union Canal, lies the 0.9ha Camley Street Natural Park. Posters and graffiti on the palings leave no one in any doubt that this is a threatened place. It is proposed as part of the terminus for the new rail route to the tunnel linking France and the UK beneath the English Channel, to be developed by the ironically named 'London Regeneration Consortium PLC'.
>
> Inside the palings, beyond the wooden chalet, with its little chairs for children and walls covered with wildlife posters and information, there lie flowery banks, developing woodland and a large pond with reed, iris and teeming insect life. Over 5000 schoolchildren, mainly from the inner city, visit this special place each year to experience a tiny bit of wild nature, often for the first time.
>
> Camley Street Natural Park was developed from an abandoned coalyard by a band of volunteers and staff of the London Wildlife Trust in 1983. The small site was landscaped, the pond dug and some species introduced to give shape and shelter. Most plants and animals, however, have arrived naturally and now there are more than 150 species of flowering plant, 18 fungi and even 70–80 species of spider
>
> (Barkham, 1995, p.81)

Figure 6.18 A wildlife gardening leaflet from *The Wildlife Trusts*. In the UK, domestic and public gardens can be a refuge for wildlife and biodiversity hotspots.

### Activity 6.6

Try to find out if there are local schemes near to where you live that attempt to use space imaginatively for biodiversity enhancement. Think about your attitude to these – are they small fry in a world of rampant degradation or are they the seeds of alternative and sustainable living? You might also consider the extent to which the term biodiversity is important to these schemes – does it capture people's imagination and values? You might also consider the role of private gardens in enhancing biodiversity.

Schemes like Camley Street Natural Park (see Figures 6.19 and 6.20 overleaf) have not escaped the attention of land developers who are keen to find ways of mitigating their building projects by finding win-win means of development (and so secure planning permission). The discourse of ecological modernization is very much apparent in cities and in other planning environments where

environmental planning
gain

For a definition of
environmental mitigation,
see Chapter Five.

developers and planners try to offset environmental costs with some procurement of environment benefit. Sometimes the provision of environmental benefit is actually specified in the planning permission. **Environmental planning gain** is a process whereby developers are 'taxed' for the increase in land value once planning permission is approved on a section of land. In return for planning permission the developer has to agree to provide some form of environmental improvement, be it a children's playground, a cycle way, or some form of landscaping and habitat improvement. This is also sometimes known as environmental mitigation. In other cases the relationship is taken a stage further with local authorities and state governments part funding a public development in partnership with private interests in a private finance initiative (PFI). These can also result in environmental mitigation measures becoming part of the agreement. While schemes vary, it is fair to say that mitigation is often viewed as rather piecemeal and tokenistic by environmentalists, and there is often disagreement over the extent to which the environmental 'improvements' or gains actually make up for the environmental costs of development. This is in part because of the rather half-hearted and unimaginative nature of many mitigation schemes. Even where biodiversity is taken more seriously, there are always uncertainties as to how successful mitigation will be in terms of what level of disturbance plants and animals will tolerate. Other questions include: will the site be used by people and non-humans in ways envisaged in the plans? And, how might changes in hydrology and microclimate affect the scheme? In other words, despite the win-win rhetoric of the ecological modernizers, biodiversity enhancement is an uncertain and experimental science, where it is impossible to say beforehand whether or not the development's benefits will balance the costs.

**Figure 6.19** An aerial view of London, showing Camley street natural park.

Local land schemes are as plagued by risks and uncertainties as those that relate to the global prospecting of genetic and plant materials. In other words, biodiversity like many environmental issues, is an intensely controversial topic where simple answers are rarely sufficient. We are in the realm then of politics, where disagreements, contests and conflicts are bound to arise that cannot be resolved simply through more knowledge.

**Figure 6.20** How the Camley street natural park looks to its visitors.

So, in answer to the question of whether or not biodiversity has a future, we can generate answers that are both optimistic (we value, understand and can incorporate more biodiversity into our developments) and pessimistic (we know so little, we only value things in ways that are marketable and therefore likely to be used up and we incorporate biodiversity in ways that are symbolic and not much more).

Looking back over some of the arguments put forward in this chapter we can fit many of the ways in which biodiversity has been handled under optimistic and pessimistic categories.

## Activity 6.7

The second column in the table below gives some general reasons why people might be, in turn, optimistic and pessimistic over future environmental prospects. Read through these reasons. Try to apply these general reasons to biodiversity specifically. Try to fill in the blanks in the right-hand column. To start you off, you have been given one reason for the pessimistic argument and one for the optimistic argument.

|  | General arguments | Arguments used in the biodiversity case |
|---|---|---|
| Optimists | Improved knowledge and know-how.<br>Greater scientific and public awareness of environmental interdependencies and of responsibilities for coexisting and future populations.<br>New institutions and regimes.<br>Political mobilization.<br>Improved communications.<br>Greater co-operation. | New biological knowledge, including genetics, has helped to raise awareness of the role of biodiversity in providing raw materials for food, medicine and roles of habitat diversity in ecosystem services.<br>... |
| Pessimists | Huge and growing disparities between the haves and the have-nots.<br>The short-term horizons of markets and governments.<br>Exploitative value systems and consumerist practices.<br>The uneven power relations that exist between states, and between corporations and people.<br>The tokenism of much environmental policy.<br>The continuing faith in technological fixes as opposed to more radical social change.<br>The tendency for environmental issues to remain distant from people's everyday lives. | Biodiversity is generally being appropriated by the North at the expense of the South.<br>...<br>...<br>... |

Your table may well look different from the one found at the end of the chapter. It should be said that this table is not too optimistic about the future of biodiversity. Things may be a little more hopeful at the local level, but the challenge for this and other areas of environmental politics is the ability to link together some of the local achievements so that they start to have the desired effect over larger areas and for longer periods.

## Summary

In this final section we have suggested that, even though the terms may differ, the biodiversity or nature on people's doorsteps may be of great value to them. Building on this level of interest in people's everyday lives is a major challenge given the global tendencies towards cultural and bio-uniformity. Finding the right language and means to enable people to engage with their environments is also a major challenge. We have also set up optimistic and pessimistic arguments regarding the future of biodiversity, which can be applied to many other topics.

# 7   Conclusion

We set out five aims in the Introduction to this chapter. They were: to identify the constraints and opportunities for sustainable development; to identify the importance of inequalities as one of those constraints; to describe scientific, economic and political risks and uncertainties with respect to the future conservation of biodiversity; to judge the extent to which ecological moderniza-tion can deliver sustainable futures; and to set up optimistic and pessimistic arguments regarding environmental futures. By now you should be able to list several opportunities and constraints, particularly with respect to the example of biodiversity. You should also be able to think more conceptually about sustainable development following the discussion of ecological modernization.

The coining of the term biodiversity has been described as an intellectual and practical response to an environmental issue (species and habitat extinctions), and the re-valuation of biodiversity in monetary terms has been discussed as a potential opportunity for the sustainable use of biodiversity. Several scientific, economic and political uncertainties regarding the measurement, valuation and conservation of biodiversity have been raised. In Section 5 we learned of the durable inequalities and power relations that make any simple solution to an environmental issue improbable. Likewise, in our critique of the discourse of ecological modernization we have suggested that environmental issues are not easily solved – the relations that have produced many of our environmental problems are not easily displaced by appeals to technology, political moderniza-tion or other reforms.

While our focus has largely been on biodiversity, you should appreciate that much of what has been said can apply to a number of environmental issues. There are many issues relating to the future of our planet and all its inhabitants about which we can be both optimistic and pessimistic. The table in Activity 6.7 can, with appropriate modification, be extended to cover issues as diverse as global climate change, waste management, transport, desertification and so on. You may have decided that our biodiversity case, and the other cases that you have come across in this book and elsewhere in the series, leaves you more pessimistic than optimistic or vice versa. Whatever your disposition, the real challenge may be to remain aware of the real complexities involved in environmental responses at the same time as looking for possibilities to increase the wellbeing of current and future generations. This is a task that cannot be realized unless the wellbeing of the world's numerous environments are also given due consideration. Whether we can meet the challenge and make the changes that are necessary to ensure a sustainable future is the question considered in the Epilogue.

# References

Andersen, M.S. (1998) (ed.) *Environmental Policy and the Role of Foreign Assistance in Central and Eastern Europe*, Denmark, KPMG/CESAM.

Balmford, A., Bruner, A., Cooper, P., Costanza, R., Farber, S., Green, R.R., Jenkins, M., Jefferiss, P., Jessamy, V., Madden, J., Munro, K., Myers, N., Naeem, S., Paavola., J., Rayment, M., Rosendo, S., Roughgarden, J., Trumper, K. and Turner, K. (2002) 'Economic reasons for conserving wild nature', *Science*, vol.297, pp.950–3.

Barkham, J. (1995) 'Ecosystem management and environmental ethics' in O'Riordan, T. (ed.) *Environmental Science for Environmental Management*, London, Longman.

**Belshaw, C.D. (2003) 'Landscape, wilderness and parks' in Bingham, N. et al. (eds).**

**Bingham, N., Blowers, A.T. and Belshaw, C.D. (eds) (2003) *Contested Environments*, Chichester, John Wiley & Sons/The Open University (Book 3 in this series).**

**Bingham, N. (2003) 'Food fights: on power, contest and GM' in Bingham, N. et al. (eds).**

**Bingham, N. and Blackmore, R. (2003) 'What to do? How risk and uncertainty affect environmental responses' in Hinchliffe, S.J. et al. (eds).**

**Blackmore, R. and Barratt, R. (2003) 'The dynamic atmosphere' in Morris, R. M. et al. (eds).**

**Blowers, A.T. and Smith, S.G. (2003) 'Introducing environmental issues: the environment of an estuary' in Hinchliffe, S.J. et al. (eds).**

**Blowers, A.T. and Elliot D.A. (2003) 'Power in the land: conflicts over energy and the environment' in Bingham, N. et al. (eds).**

**Burgess, J. (2003) 'Environmental values in environmental decision making' in Bingham, N. et al. (eds).**

**Castree, N.C. (2003) 'Uneven development, globalization and environmental change' in Morris, R.M. et al. (eds).**

Costanza, R., d'Arge, R., de Groot, R.S., Farber, S., Grasso, N. and Hannon, B. (1997) 'The value of the world's ecosystem services and natural capital', *Nature*, vol.387, pp.253–60.

Cronon, W. (1996) 'The trouble with wilderness' in Cronon, W. (ed.) *Uncommon Ground: Rethinking the Human Place in Nature,* New York, Norton.

**Drake, M. and Freeland, J.R. (2003) 'Population change and environmental change' in Morris, R.M. et al. (eds).**

Ford-Lloyd, B. and Jackson, M. (1986) *Plant Genetic Resources*, London, Edward Arnold.

**Freeland, J.R. (2003) 'Are too many species going extinct? Environmental change in time and space' in Hinchliffe, S.J. et al. (eds).**

GRAIN (Genetic Resources Action International) (2002) 'Growing diversity', *Seedling*, July 2002, pp.16–21.

**Hinchliffe, S.J. and Belshaw, C.D. (2003) 'Who cares? Values, power and action in environmental contests' in Hinchliffe, S.J. et al. (eds).**

**Hinchliffe, S.J., Blowers, A.T. and Freeland, J.R. (eds) (2003) *Understanding Environmental Issues*, Chichester, John Wiley & Sons/The Open University (Book 1 in this series).**

Leopold, A. (1948) *A Sand County Almanac, and Sketches Here and There*, Oxford, Oxford University Press.

Mol, A. (1995) *The Refinement of Production,* Utrecht, van Arkel.

**Morris, R.M. (2003a) 'Changing land' in Morris, R.M. et al. (eds).**

**Morris, R.M. (2003b) 'General conclusions: thinking about environmental change' in Morris, R.M. et al. (eds).**

**Morris, R.M., Freeland, J.R., Hinchliffe, S.J. and Smith, S.G. (eds) (2003) *Changing Environments*, Chichester, John Wiley & Sons/The Open University (Book 2 in this series).**

**Morris, R.M. and Turner, C. (2003) 'Dynamic Earth: processes of change' in Morris, R.M. et al. (eds).**

Pearce, D. and Moran, D. (1994) *The Economic Value of Biodiversity*, London, Earthscan/IUCN.

**Reddish, A. (2003) 'Dynamic Earth: human impacts' in Morris, R.M. et al. (eds).**

Shiva, V. (1995) 'Biotechnological development and the conservation of biodiversity' in Shiva, V. and Moser, I. (eds) *Biopolitics: a Feminist and Ecological Reader on Biotechnology*, London, Zed Books, pp.193–213.

Wilson, E.O. (1992) *The Diversity of Life*, London, Penguin Books.

Wilson, E.O. and Peter, F.M. (1988) *Biodiversity*, Washington, National Academy Press.

Yaron, G. (2001) 'Forest, plantation crops or small-scale agriculture? An economic analysis of alternative land use options in the Mount Cameroon area', *Journal of Environmental Planning and Management*, vol.44, no.1, pp.85–108.

# Answers to Activities

## Activity 6.2

| Monocultural forest | Diverse forest |
| --- | --- |
| timber | genetic pool |
| fuelwood | non-wood products |
| other business products | agricultural production |
| | recreation and tourism |
| | recycling nutrients |
| | watershed protection |
| | protecting soil quality and erosion resistance |
| | habitat for people, flora and fauna |
| | aesthetic, cultural and spiritual source |
| | scientific data |

## Activity 6.7

| | General arguments | Arguments used in the biodiversity case |
| --- | --- | --- |
| optimists | Improved knowledge and know-how. Greater scientific and public awareness of environmental interdependencies and of responsibilities for coexisting and future populations. New institutions and regimes. Political mobilization. Improved communications. Greater co-operation. | New biological knowledge, including genetics, has helped to raise awareness of the role of biodiversity in providing raw materials for food. Biodiversity conservation is backed up by a convention (Convention on Biodiversity, CBD) and national and local action plans (Biodiversity Action Plans, BAPs) – indicating the rise of new institutions, global cooperation and communication. |
| pessimists | Huge and growing disparities between the haves and the have-nots. The short-term horizons of markets and governments. Exploitative value systems and consumerist practices. The uneven power relations that exist between states, and between corporations and people. The tokenism of much environmental policy. The continuing faith in technological fixes as opposed to more radical social change. The tendency for environmental issues to remain distant from people's everyday lives. | Biodiversity is generally being appropriated by the North at the expense of the South. Biodiversity prospectors are set on making short- to medium-term profits, and lack a sensitivity to loss of knowledge and skills. Even though relative biodiversity is high in poorer countries, large north American and European companies hold most of the economic, legal and scientific cards, which may result in most value being appropriated elsewhere. The Convention on Biodiversity is largely a toothless agreement. Biotechnology is seen as an answer to biodiversity loss, but a more radical overhaul of power inequalities is needed. 'Biodiversity' may not mean a great deal to people involved in conservation issues – other terms may be better at capturing people's experiences of environments. |

# Epilogue: environmental futures

Andrew Blowers and Steve Hinchliffe

## A moment of crisis?

At the conclusion of his account of the twentieth century the historian Eric Hobsbawm wrote:

> The future cannot be a continuation of the past, and there are signs, both externally, and, as it were, internally, that we have reached a point of historic crisis. The forces generated by the techno-scientific economy are now great enough to destroy the environment, that is to say, the material foundation of human life. The structures of human society themselves, including even some of the social foundations of the capitalist economy, are on the point of being destroyed by the erosion of what we have inherited from the human past. Our world risks both explosion and implosion. It must change.

(Hobsbawm, 1995, pp.584–5)

This apocalyptic pronouncement reveals a deep pessimism born of careful analysis and awareness of the risks that human actions are imposing on environmental security. Such a pessimistic portrayal of environmental futures counters the sanguine belief in the possibilities of achieving sustainable development through ecological modernization that we discussed in Chapter Six. The technological, economic and political approaches explored in earlier chapters have had many beneficial environmental impacts. They have contributed to a cleaner environment, encouraged conservation, and improved public health and well-being especially on local and regional scales. Although these benefits have been unevenly distributed they indicate a developing response to environmental change achieved within existing social structures and institutions. Nevertheless, the environmental risks the world now faces from climate change, radioactivity and biotechnology outlined in this book suggest the need for a social response that goes beyond a broad continuation of existing trends. In Hobsbawm's words, our world must change.

Hobsbawm focuses on the contemporary human world; his is an anthropocentric concern. In so far as humans have created the problems, it is for humans to discover and implement the solutions. In what sense can it be said we have reached a 'historic crisis' in terms of the relationship between environment and society? We need to be cautious here for every society in every epoch has faced great uncertainty and sometimes risks that may have seemed every bit as threatening as those we face today. Consider the threats from disease, famine, warfare and destruction that have occurred throughout human history and that are still routinely encountered in some parts of the world. But there are, perhaps, two differences. First, there is the change in the scale of the risks the world now runs. Second, there is a change in the degree to which we feel we have a choice regarding different risks.

Time and space have been used as key analytical concepts throughout this series, notably in *Changing Environments* (**Morris et al. (eds), 2003**).

The question of increasing scale is a defining feature of the present. It is worth noting how our sense of time and space compares with earlier periods. In the Middle Ages, for example, time was not conceived in terms of a fixed and continuous timetable with precise dates and demarcations of activity. Rather, as the detailed portrait of the Pyrenean community of Montaillou reveals, in rural society time was often fixed by reference to natural events such as harvests or religious festivals; it was vague and people had a relaxed, unregulated attitude to the passing of time and there was an absence of a historical dimension (Ladurie, 1978). In terms of space, the focus was the village. The locality and the sub-region were demarcated by natural obstacles such as mountains and rivers. There were wider connections made through pilgrimage and trade. The merchants of the Middle Ages developed extensive connections through mountain passes, along rivers and main highways, focusing on key centres such as the Champagne Fairs of the thirteenth century and the emerging industrialized areas of Flanders, northern Italy and southern Germany and the growing cities of Paris and London (Spufford, 2002). These European networks were also linked to areas much further afield in China and India (Figure 1). Along these routes were carried goods, ideas, art, technologies and disease promoting industrial, cultural and environmental change. Thus the wider spatial horizons, which characterized the development of modernization, began to emerge.

**Figure 1**  Trade in the Gulf of Cambay, India, by Boucicaut Master (fl. 1390–1430) from *Livres des Merveilles du Monde* (c.1410–1412).

In our own time these horizons have become limits. The risks that attend the increase and spread of pollution or the depletion of resources and biodiversity are not confined: they spread inexorably until they include ever larger parts of lands, oceans and atmosphere and in some cases, such as climate change, can even be said to affect virtually the whole planet simultaneously. At the same time the risks arising from contemporary activities extend down the generations. As we saw in Chapter Five, the effects of global climate change will be felt by future generations whatever action is taken to reduce greenhouse gas emissions now. Chapter One emphasized that the hazards from radioactivity will persist over unimaginably long timescales irrespective of how we manage the legacy of radioactive wastes. And in Chapter Six we saw the possibilities of modern biotechnologies transforming environments from biodiversity to bio-uniformity.

The second possibly distinct aspect of our present time relates to choices – however limited they may be – over our futures. In the past the hazards of living were perhaps less often perceived of as risks that might be minimized. It was often the case the hazards were faced as chance events over which people had little or no control (apart perhaps from consulting oracles or conforming to the wishes of a deity). In this sense societies were often fatalistic, assuming that what will be will be and surrendering to a god's will. In our present times, while it would be wrong to say that such fatalism or beliefs have necessarily diminished everywhere, there is a sense that risks can be evaluated and to an extent managed. So in the chapters of this book there has often been an implicit optimistic message that through the application of better decision-making procedures, technologies, economic instruments, national and international political structures and forms of modernization, risks could be reduced in the future. This sense of having some degree of influence on the future puts an extra burden on societies, governments, business and other organizations to take some responsibility for those futures.

We may conclude that the present epoch is distinguished by risks on a new scale, and a growing sense of responsibility for managing those risks.

# Prospects for change

The idea that we have reached a moment of crisis and that our world must change is at once depressing and stimulating. It urges us to consider the question we posed at the beginning of this book: what kinds of environmental futures are desirable and how might we choose among them? To answer this we have to understand what opportunities exist for change and what constrains the freedom to choose. What kinds of change are necessary and what shifts can we perceive to be occurring? These questions cause us to reflect on another set of analytical concepts, the concepts of values, power and action.

The analytical concepts of values, power and action have been used throughout this series, especially in *Contested Environments* (Bingham et al.( eds), 2003).

# Changing values

With regard to the concept of values what emerges from this book is that environmental responses seem predominantly instrumental. Responses, at a technical or political level, are directed towards minimizing environmental risk without infringing economic development. There is a concern with conservation to save resources and costs. There is an emphasis on controlling and reducing pollution and waste to enhance environmental amenity and improve health. In richer societies environmental quality becomes a component of improving quality of life. The predominant value system is utilitarian, emphasizing the role of environments in contributing to human welfare. In Chapters Three and Six we saw how there is a tendency to try to represent these environmental values in monetary terms, to evaluate them in terms of economic criteria. This materialist approach sets out environmental futures in terms of a 'weak' form of sustainable development. This form of sustainability may be realized through ecological modernization that encourages the application of green technologies, the implementation of economic instruments and the encouragement of partnership between the state and the private sector to ensure a balance between economic and environmental objectives.

In Chapter Three the limitations of a utilitarian approach were discussed. In particular utilitarianism is predominantly concerned with the present rather than the future, with the human rather than the nonhuman world, with economic rather than non-economic measures of happiness or satisfaction. Alternative value systems may reflect a much stronger form of sustainability, one which takes into account the rights of future generations and nonhuman aspects of environments and which recognizes that satisfaction relates to such things as opportunity for leisure, contribution to society, greater personal autonomy and so on. Again, this reflects the choices available to more affluent societies but, in this case, the emphasis is on well-being rather than wealth. Taken to its conclusion this alternative value system counters the assumptions of ecological modernization that economic growth and environmental conservation are mutually consistent aims.

Unlike ecological modernization, which emphasizes growth and continuity, values based on strong interpretations of sustainable development imply considerable shifts in behaviour. Instead of an emphasis on individual consumerism there would be a commitment to the provision of collective (and often more environmentally friendly) provision such as public transport, to forms of food production that do not endanger humans and nonhumans, to renewable energy technologies that conserve resources and to avoidance of waste and activities that increase the risks of long-term pollution hazards. While such shifts are widely advocated and partially achieved their fulfilment would represent a major shift in the way we live. They would probably require behavioural changes that most people would not contemplate. They imply priority for the common good that does not resonate with an age of individualism. Such changes are unlikely to be embraced in neo-liberal democracies where short-term political success depends primarily on providing economic stability and growth. Unfortunately the prospects for a fundamental shift in behaviour, which Chapter Five indicates to be necessary to respond to global climate change, are only likely to increase once the deleterious environmental consequences are unmistakable and irreversible. By then it may be too late.

## Changing power relations

Environmental futures also depend on power relations. Power acts as a constraint on change, and yet changing power relations and discourses can also be vehicles whereby change is made possible. This book provides ample evidence of power as a constraint. The concentration of power resources in the hands of experts, in large business corporations and in political elites constitutes a formidable obstacle to alternative conceptions of environmental futures. The development of discourses of modernization – that seem to offer win–win solutions to environmental problems, but that may well disguise a business-as-usual approach – similarly obscure the workings of powerful forces.

The inequality of power was especially obvious in the reactions to the Chernobyl nuclear fall-out discussed in Chapter One where expert views backed by government were imposed on sheep farmers with severe consequences for the farmers. In Chapter Four we saw examples of domination over environmental policy making by business and government that were sufficient to suppress or block alternative conceptions of environmental futures. A sense of powerlessness was experienced by the Cumbrian hill farmers, by the mining community of Aznalcóllar and by communities along the Betuwe railway route. In this book we have tended to focus on inequalities of power within wealthy nations especially in western Europe. Obviously, inequalities of power are also a condition of the relationships between North and South. In the South powerlessness and poverty often combine to present levels of environmental degradation and risk that would not be tolerated in the North. The focus in Chapter Six on the power imbalances between agri-industrial companies and farmers in developing countries serves as an example. Meanwhile, we would argue that these imbalances are starting to emerge strongly in the relationship between the developed nations and the countries in transition whose resources are increasingly being opened up to large-scale exploitation (Chapter Six).

The inequality of risk between North and South can be illustrated by the world's biggest industrial accident so far which occurred in 1984 in the city of Bhopal, central India. Union Carbide, an American multinational chemical company, had constructed a plant to produce Sevin, a pesticide that would control the insects that ruined crops across the sub-continent. One of the chemicals used was methyl isocyanate (MIC) which, if inhaled, has a deadly impact on the respiratory

**Figure 2** The disaster at Bhopal, December, 1984: (left) a woman from a nearby slum collects water in front of the Union Carbide plant; and (right) apart from the thousands killed, huge numbers of people suffered nausea, shortness of breath and red eyes.

system. Although the risks from the deadly gas were well known, MIC was stored in tanks at the plant, a practice quite contrary to that in equivalent plants in the USA. With production of Sevin declining in the early 1980s the plant operated intermittently and, when production ceased, some of its primary safety systems were shut down. On the night of 3 December a leak occurred in the MIC tanks that could not be controlled and led to the escape of deadly clouds of MIC gas, which enveloped nearby *bustees*, poor neighbourhoods lived in by many of those formerly employed by the company. Estimates of casualties vary, ranging from 3,000 to 30,000 deaths and up to half a million suffering from the effects of the toxic cloud (Lapierre and Moro, 2002). The case illustrates how social inequality is reflected in inequalities of risk. The Union Carbide plant in India was able to operate without all the strict regulations applied to equivalent plants in the USA. The workers at the plant were recruited from among the very poor across India and most lived in the *bustees*. It was this population that suffered the brunt of the impact. It was an environmental disaster that discriminated against the poor (Figure 2).

Bhopal, and other serious environmental accidents such as Seveso, Basel and Chernobyl were important, as Chapter Four shows, in gaining attention for environmental concerns. They have also stimulated a challenge to the established power relations between experts, business and science. Bhopal demonstrates how faith and trust were initially invested in a technology which promised environmental and economic benefits in the form of better crop production, jobs and wealth. The accident revealed how that trust was betrayed by a multinational which was intent on cutting costs at the expense of safety. Trust in government was also forfeited by the inability of politicians to ensure safe operation and adequate compensation for the victims of Bhopal.

Bhopal provides an instance of a growing breakdown in trust in expertise (or, if trust was never that abundant, Bhopal reflects a growing divide between politicians, big business and the public, one that is unlikely to be bridged by trust in the future). In the wider context this partly reflects a more sceptical, less deferential age. But the loss of trust also reflects a social response to the evident failure of experts, governments and businesses to avoid risks and ensure the safe management of dangerous technologies. This has helped to shift the balance of power towards environmental movements. Nonetheless the shift in power relations may not yet have proceeded very far, even in the wealthier countries. Ranged against the formidable resources of big business, governments and a consumer orientated society the countervailing resources of those concerned to protect environments seem pusillanimous indeed. What we can say is that the contest is becoming a little less uneven.

## Action and change

The shift in power that is occurring has brought results through environmental actions. In terms of both practical policy making and new ways of thinking, there are grounds for a more optimistic outlook on our environmental futures. Indeed, there are many instances in the developed world, observed day by day, of species protected, fish returning to erstwhile polluted rivers, degraded environments restored, urban wildlife parks created, air quality improved, dumping in oceans banned and so on. Chapter Two gives a range of examples whereby new technologies are being introduced to improve methods of waste management, conserve energy, minimize pollution and control congestion. It may not seem to amount to much against the loss of hedgerows, ancient forests or marshlands, the declining numbers of once common birds, the stockpiles of wastes, the oil slicks on the shorelines, the contaminated water or smog in the cities. Elsewhere, in many parts of the developing world and, as we have seen, in countries in transition such as Russia environmental degradation is widespread and severe. But there is also evidence of improvement in some areas; for example, the efforts to protect Lake Baikal indicate that environmental conservation is gaining ground where important habitats are threatened.

New ways of thinking about environments are being stimulated through the discourse of 'sustainable development', which has become pervasive in environmental policy making. It has entered political rhetoric at all levels. The environment has become a key political and moral issue as the White Paper, *This Common Inheritance* proclaimed in 1990: 'We have a moral duty to look after our planet and to hand it on in good order to future generations. That is what experts mean when they talk of "sustainable development": not sacrificing tomorrow's prospects for a largely illusory gain today' (HMSO, 1992, p.10).

Of course, it will take more than routine rhetorical pronouncements to achieve the shifts in behaviour and changes in power relations necessary to make sustainable development more of a reality. But a start has been made. The importance of statements such as the one above is that it requires sustainability to be built into policy thereby making policy makers consider environmental risks and consequences over a longer term. It is, perhaps, at this level of considering the implications of our actions that the best hope for our environmental futures lies (Figure 3).

Figure 3 Hope for future environments? Urban cultivation projects like this one in east Birmingham called 'concrete to coriander', run by the Community Service Volunteers (CSV), are starting to green our cities and raise awareness of environmental issues. The images show a derelict site before and after cultivation.

# The components of change

The kind of changes in values and power relations necessary to ensure sustainable forms of development can only be brought about by widening access to knowledge and understanding. From the discussions in this book it can be concluded that these changes can be made effective through three social processes. One is greater **openness**, which implies a cultural change by opening up debate and policy making to scrutiny by those with different values, agendas and objectives. Another, closely related, is greater **participation**, which means providing greater equality of access and involvement in decision making. A third is greater **dissension**, a willingness to challenge received wisdoms, to protest at injustices and to change political processes from the outside. The evidence from this book suggests there is a long way to go but there are also some signs of emerging changes.

openness

participation

dissension

## Openness

Lack of knowledge and of understanding of environmental processes and of the social implications is evident in the cases explored in this book. Ignorance is not confined to those who are excluded from debate. As the Cumbrian case shows (Chapter One) scientific experts' claims to knowledge can be confounded by empirical evidence that their methods fail to take into account. It was evident that scientific experts and government officials proceeded on the basis of applying a theory developed in relation to soils in one area to the quite different circumstances of the Lakeland fells. Their field of theoretical knowledge did not encompass the practical and intuitive experience of sheep farmers whose livelihoods depended on understanding their local environment.

In any case, experts do not always agree. Expert knowledge may be contested by other experts resulting in a failure to achieve scientific consensus, as the Sellafield RCF case illustrates. The case also shows that expert judgements may be contaminated by political exigencies, the need for speedy decisions or to avoid excessive costs. There is often the need to reach necessary compromises to satisfy conflicting interests. At another level the complex negotiations over climate change discussed in Chapter Five provide ample evidence of the desire to reach negotiated solutions through compromise.

There has been a tendency for experts to have privileged access to decision makers and to be protected by closed and secretive decision making. More and more this elitist and exclusive decision making has been challenged by environmental groups and by ordinary citizens opposed to specific projects and able to mobilize counter expertise and action. They have made the case for openness and transparency of processes of decision making. Put into practice this means open peer review of scientific evidence, sufficient time to ensure that all appropriate evidence can be considered and efforts to ensure that knowledge is made widely accessible. In short, knowledge about environmental processes and their consequences must be democratized.

## Participation

Whereas greater openness is an enabling process, widening participation is an empowering process. There has been some progress in this area, too. Policy formation has gone beyond the 'Decide Announce Defend' practice which so often resulted in failure. And the process of formal consultation where proposals are set out and responses taken into account before a decision is made has also proved inadequate in gaining public acceptability. More and more, participation in helping to form policy has been encouraged. Environmental movements have become more involved in stakeholder dialogues as members of advisory bodies, and they may also be involved in the implementation of policies. At the international level environmental NGOs have become increasingly active in seeking to influence regimes in such areas as climate change, protection of biodiversity and prevention of desertification.

This participation, though widening, is itself also exclusive: it is confined to experts, scientists and other so-called 'stakeholders' who, by definition, have a particular interest in an issue. Among these may be NGOs who have achieved access to decision making but are not necessarily representative of the interests of society as a whole. As Chapter Four has demonstrated, for all the emphasis on openness and participation in environmental policy, power remains concentrated in the hands of business, government and self-selected interests drawn from civil society. The power of representative institutions (Parliament and local government) has diminished and substantial sections of society have little or no involvement or influence over many areas of environmental decision making.

The lack of openness and participation leads to confrontation and, increasingly, there is resistance to projects and policies. There is a need to engage with a wider public to achieve understanding and acceptability of proposals. New forms of policy making are needed that reinvigorate democratic processes upon which consent is ultimately based. They can combine a greater openness, which reveals the nature of problems, and a greater participation, which widens the basis for the acceptability of solutions. As we write, there are the first tentative signs that this may be happening. In just one field of policy, radioactive waste management (Chapter One), the UK has embarked on an innovatory and experimental process of dialogue and public involvement aimed at identifying an option or combination of options for managing radioactive waste that will achieve long-term protection for people and the environment. This approach is distinctive in two respects. One is that it actively seeks public involvement through a wide range of techniques designed to provide information, gather opinion, improve understanding, engage in dialogue and deliberation and establish criteria for assessing options. The effort to secure public support goes far beyond anything hitherto attempted. The process is also distinctive in that it transcends the normal cycle of politics. Radioactive waste is a classic instance of an issue that does not fit electoral timescales. Decisions taken now may have implications spanning generations into the far future. With its emphasis on openness, on public involvement and on the longer term, the process is intended to gain public acceptability and hence legitimacy for policies that must survive short-term political and economic pressures.

This is one example in one specific field of policy but it represents new ways of thinking about environmental futures. Such approaches should be seen as complementary to and not as a substitute for democratic processes. They are a way of opening up policy making. In so doing they wrest at least some power from business, government and experts and broaden the constituency and scope of policy making.

## Dissension

Making policy so open and accommodating risks losing sight of the fact that some of the big questions do not fit neatly into a re-jigged political process. Indeed, the more participation is fostered the more there is a danger of formerly dissenting voices becoming co-opted into the larger scheme of things. There are also some very real limits on the extent to which people who lead busy lives can spend large amounts of time in formal political discussions. For these and other reasons, we should not overlook the fact that protest and direct action are becoming more and more common and perhaps necessary in our times (Figure 4). Environmentalism itself is starting to look quite divided with so-called professional environmentalists working within the system, so to speak, and a raft of others protesting on any number of issues. Environmentalism has in more recent times become a plural movement. Environmental radicals, eco-feminists, direct activists and environmental justice campaigners and others see attending formal meetings as only part of their job. The other part is to try to challenge

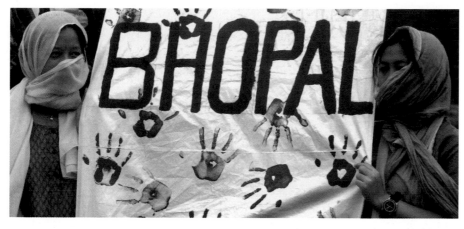

Figure 4 Signs of change. Indian activists protest during the ongoing criminal case on the disaster in Bhopal, at a rally in Bombay, India, 2002.

society's norms and discourses head on and to offer up more radical alternatives to our current ways of life.

The complexity of environmental processes makes changes in policy responses – and perhaps in whole ways of life – essential. The world, in Hobsbawm's words, must change if it is to avoid destruction. There is a need for new institutions to deal with the transboundary and transgenerational nature of environmental problems. These are beginning to emerge at national level, indicated by the example of radioactive waste, and at international level, where regimes are developing to grapple with global environmental problems. At the same time, these advances cannot neglect the local conditions in which people live. There is growing awareness of the major challenge that faces us: to try to find ways of living sustainably while overcoming some of the constraints in finding practical solutions to what are currently unsustainable lifestyles. Those are, perhaps, the early signs that the world is beginning to change. They reflect the hope that we can respond effectively, that the direction of change is not irreversible and that we have a choice over what we do.

# References

Bingham, N., Blowers, A.T. and Belshaw, C.D. (eds) (2003) *Contested Environments*, Chichester, John Wiley & Sons/The Open University (Book 3 in this series)

HMSO (1992) *This Common Inheritance,* London, CM 1200.

Hobsbawm, E. (1995) *Age of Extremes*, London, Abacus.

Ladurie, E. Le Roy (1978) *Montaillou,* Scolar Press (published by Penguin Press, Harmondsworth, 1980 and reprinted in Penguin and Peregrine Books).

Lapierre, D. and Moro, J. (2002) *Five Past Midnight in Bhopal,* London, Scribner.

Morris, R.M., Freeland, J.R., Hinchliffe, S.J. and Smith, S.G. (eds) (2003) *Changing Environments*, Chichester, John Wiley & Sons/The Open University (Book 2 in this series).

Spufford, P. (2002) *Power and Profit: The Merchant in Medieval Europe,* London, Thames and Hudson.

# Acknowledgements

Grateful acknowledgement is made to the following sources for permission to reproduce material within this book.

## Figures

*Chapter One contents page and Figure 1.6:* © Tropix Picture Library; *Figure 1.1b:* © Chris Demetriou/Frank Lane Picture Agency; *Figure 1.2:* Gould, P. (1990) *Fire in the Rain: The Democratic Consequences of Chernobyl*, Blackwell Publishing Ltd, Copyright © 1990 Peter Gould; *Figure 1.3:* 25th Scientific Report of the Institute of Terrestrial Ecology (1986). With permission from the Centre for Ecology and Hydrology; *Figure 1.4:* NRPB Report 191, National Radiological Protection Board; *Figure 1.5: Post Closure Performance Assessment: Treatment of the Biosphere*, Report No. S/95/002, Nirex Ltd; *Figure 1.11:* © British Nuclear Fuels Ltd; *Figure 1.8a:* © Mike Dodd; *Figure 1.8b:* © Cranfield University, Silsoe; *Figure 1.8c:* Extract from Ordnance Survey map of Birkhouse Moor, Cumbria, with the permission of Ordnance Survey on behalf of The Controller of Her Majesty's Stationery Office, © Crown copyright, licence number ED100020607; *Figure 1.9:* © John Watkins/Frank Lane Picture Agency; *Figure 1.10: Going Forward*, Nirex Ltd; *Figure 1.12:* © John Giles/PA Photos; *Figure 1.13a:* Crown copyright material is reproduced under Class Licence Number C01W0000065 with the permission of the controller of HMSO and the Queen's Printer for Scotland; *Figures 1.13b and 1.14b*: Courtesy of Nirex Ltd; *Figure 1.15:* Nirex Report 71, 1989 and Nirex Report 263, 1992, Nirex Ltd; *pages 46 and 48:* Nilsson, A. (2001) *Responsibility, Equity and Credibility*, Kommentus Forlag.

*Chapter Two contents page and Figures 2.1a, 2.1b, 2.31a, 2.31b and 2.31c:* © NASA/Goddard Space Flight Center Scientific Visualisation Studio; *Figure 2.2:* National Geographic Magazine, November 2002, pp.76–7, National Geographic Society; *Figure 2.3:* http://www.defra.gov.uk/environment/statistics/eiyp/general/gen59.htm. Crown copyright material is reproduced under Class Licence Number C01W0000065 with the permission of the controller of HMSO and the Queen's Printer for Scotland; *Figure 2.6:* © US Environmental Protection Agency (EPA), Science Advisory Board 1990, Reducing Risk: Setting Priorities and Strategies for Environmental Protection; *Figure 2.8:* © NASA/EP TOMS; *Figure 2.11:* © Innogy plc; *Figure 2.14:* © Herbert Giradet/Still Pictures; *Figure 2.19:* Courtesy of Aylesbury Auto Salvage, Photo Mike Levers/OU; *Figure 2.20:* Photo: Rod Barratt (author); *Figure 2.22:* © Hank Morgan/Science Photo Library; *Figure 2.27:* © Courtesy of Biffa Waste Services Ltd.; *Figure 2.29:* © Photo: ENERGI E2; *Figure 2.30:* © Oldrich Karasek/Getty Images.

*Chapter Three contents page and Figure 3.3:* © Ron Giling/Still Pictures; *Figure 3.1:* © Jim James/PA Photos; *Figures 3.6, 3.7, 3.9, 3.10 and 3.12:* © Ed McLachlan; *Figure 3.11:* © Peter Oxford/Nature Picture Library;

*Figure 3.13:* (left) Mark Campbell/Link Picture Library, (right) © P. Anand/Link Picture Library; *Figure 3.14:* © Mike Keefe/Cagle Cartoons, Inc.; *Figures 3.15 and 3.16:* These figures first appeared in April 2001 issue of *The Ecologist* (vol.31, no.3) www.theecologist.org

*Chapter Four contents page and Figure 4.3a:* © Emilio Morenatti/Associated Press; *Figure 4.2a:* Photo: Ecologists in Action, Madrid; *Figures 4.2b and 4.3b:* © La Junta de Andalucia; *Figure 4.4: (left)* © David Drain/Still Pictures, *(top right)* © Michel Gunther/Still Pictures, *(bottom right)* © J.L.Gomes de Franciso/ Nature Picture Library; *Figures 4.6 and 4.8:* Pestman, P. (2001) In Het Spoor Van de Betuweroute - Mobilisatie, Besliutvorming en institutionalisering rond een groot infrstructureel project, Rozenberg Publishers, Nijmegem/Amsterdam; *Figure 4.7:* Wild Life Crossings for Roads and Waerways, Dutch Ministry of Transport, Public Works and Watcr Management, Delft, The Netherlands, Prof. Leroy; *Figure 4.9:* Weale, A. et al. (2000) Environmental Governance in Europe: An Ever Closer Ecological Union?, Oxford University Press; *Figure 4.10:* Courtesy of the EU Eco-label Helpdesk.

*Chapter Five contents page:* © Jack Whelan/ICCWBO; *Figure 5.1: (left)* © Cyril Ruoso/Still Pictures, *(top right)* © Paul Mattsson/Report Digital, *(bottom right)* © Gustavo Ferrari/Associated Press; *Figure 5.3: (left)* © H.Baesemann/UNEP/ Still Pictures, *(top right)* © Obed Zilwa/Associated Press, *(bottom right)* © Jim Foster/English Nature; *Figure 5.5:* Climate Change 2001: Mitigation. Inter-governmental Panel on Climate Change; *Figure 5.7: (left)* © Daniel O'Leary/ Panos Pictures, *(top right)* © Pieternella Pieterse/Still Pictures, *(bottom right)* © Andy Crump/WHO/TDR; *Figure 5.8: (top left)* © Scottish Power plc, *(bottom left)* © Rocky Mountain Institute, Colarado, *(bottom right)* © Tommaso Guicciardini/Science Photo Library; *Figures 5.8 (top right), 5.10, 5.12, 5.17, 5.22, 5.30, 5.31:* Photos: Stephen Peake (author); *Figure 5.14:* © International Telecommunications Union/A. de Ferron; *Figure 5.15a:* © Popperfoto; *Figure 5.15b:* © Daniel Garcia/Popperfoto; *Figure 5.15c:* © EPA/PA Photos; *Figure 5.16:* Courtesy of UNFCCC secretariat; *Figure 5.18:* From: Agarwal, A. Making the Kyoto Protocol Work, Centre for Science and Environment, India; *Figure 5.20: (left)* © AFP/Popperfoto; *Figure 5.20 (right) and 5.21:* Photos: Sarah Peake; *Figure 5.23:* © Toshiyuki Aizawa/Reuters/Popperfoto; *Figure 5.24:* Methodological Issues, National Communications from parties included in Annex 1 to the Convention. Report on national greenhouse gas inventory data from Annex 1 Parties for 1990–2000, UNFCCC; *Figure 5.25:* Toles © 1995 The Buffalo News reprinted with permission of Universal Press Syndicate. All rights reserved; *Figure 5.28:* Courtesy of the Federation of Electric Power Companies; *Figure 5.29:* © International Telecommunications Union, 2003; *Figure 5.33:* © GADO/United Nations Environment Programme.

*Chapter Six contents page and Figure 6.13b:* © Gil Moti/Still Pictures; *Figure 6.1:* Plate 3 'La Forêt du Brésil' from *Voyage Pittoresque dans le Brésil* by Johann Moritz Rugendas, 1835. Reference (shelfmark) 2096 a.1. Bodleian Library, University of Oxford; *Figures 6.2a and 6.2b:* © Peter Wakely/English Nature;

*Figures 6.3a and 6.3b:* © Mike Dodd; *Figure 6.4:* Costanza et al. (1997) 'The value of the world's ecosystem services and natural capital', *Nature*, vol.387, pp.253–60. http://www.nature.com; *Figure 6.5:* Reprinted with permission from Balmford, A. (2002) 'Economic reasons for conserving wild nature', *Science*, vol.297 © 2002 American Association for the Advancement of Science; *Figure 6.6:* © Fiacomo Pirozzi/Panos Pictures; *Figure 6.7b:* © Edward Parker/Still Pictures; *Figure 6.10a:* © Jon Spaull/Panos Pictures; *Figure 6.10b:* © Doug Allen/Nature Picture Library; *Figure 6.11:* © Oliver Gillie/Still Pictures; F*igure 6.12:* © Roland Seitre/ Still Pictures; *Figure 6.13a:* © Heidy Bradner/Panos Pictures; *Figure 6.14a:* © Michel Bureau/Still Pictures; *Figure 6.14b:* © Heidi Bradner/Still Pictures; *Figure 6.15a:* © Trygve Bolstad/Panos Pictures; *Figure 6.15b:* © Adrian Evans/ Panos Pictures; *Figure 6.17:* Courtesy BTCV, photograph © Jane Alexander, Courtesy of The Countryside Agency, 'Ecoregen' and 'Building Sustainable Communities', Groundwork UK; *Figure 6.18:* © The Wildlife Trusts; *Figures 6.19 and 6.20:* © London Wildlife Trust.

*Epilogue: Figure 1:* From Livres des Merveilles du Monde (c.1410-12) © Biblioteque Nationale, Paris, France/Bridgeman Art Library; *Figure 2: (left)* © Henderson/Associated Press, *(right)* © Sondeep/Associated Press; *Figure 3: (left and right)* © Community Service Volunteers (CSV) Environment; *Figure 4:* © Rajesh Nirgude/Associated Press.

## Tables

*Table 1.1: Post Closure Performance Assessment: Treatment of the Biosphere*, Report No S/95/002. Nirex Ltd; *Table 3.3:* Adapted from Barrett, S. (2000) *Oxford Review of Economic Policy*, Oxford University Press; *Table 6.1:* Pearce, D. and Moran, D. (1994) *The Economic Value of Biodiversity*, Earthscan Publications Ltd.

## Cover illustrations

From left to right: © London Wildlife Trust; © EPA/PA Photos; © Still Pictures/ F. Ardito; © Innogy plc.

Every effort has been made to contact copyright holders, but if any have been inadvertently overlooked the publishers will be pleased to make the necessary arrangements at the first opportunity.

# Index